DURB CURLEE

MALCOLM GAY

THE BRAIN ELECTRIC

Malcolm Gay is an arts reporter for *The Boston Globe*. His writings and essays have appeared in *The New York Times*, *The Atlantic*, and Time.com, among other publications. Named an Alicia Patterson Fellow in 2013, Gay has won numerous journalism awards.

THE BRAIN ELECTRIC

THE BRAIN ELECTRIC

THE DRAMATIC HIGH-TECH RACE TO MERGE MINDS AND MACHINES

MALCOLM GAY

FARRAR, STRAUS AND GIROUX NEW YORK

Farrar, Straus and Giroux
18 West 18th Street, New York 10011

Printed in the United States of America
Published in 2015 by Farrar, Straus and Giroux
First paperback edition, 2016

The Library of Congress has cataloged the hardcover edition as follows:
Gay, Malcolm, 1972– author.
The brain electric : the dramatic high-tech race to merge minds and machines / Malcolm Gay.
p. ; cm.
Includes bibliographical references and index.
ISBN 978-0-374-13984-1 (hardback) — ISBN 978-0-374-70962-4 (e-book)
I. Title.
[DNLM: 1. Brain-Computer Interfaces. 2. Biomedical Engineering—trends. 3. Man-Machine Systems. 4. Research Subjects. WL 26.5]

R856
610.28—dc23

2015010114

Paperback ISBN: 978-0-374-53641-1

Designed by Jonathan D. Lippincott

P1

For my parents, all four of them

Man exists only insofar as he is separated from his surroundings. The cranium is a space-traveler's helmet. Stay inside or you perish . . . It may be wonderful to mix with the landscape, but to do so is the end of the tender ego. —Vladimir Nabokov, *Pnin*

CONTENTS

THE BRAIN ELECTRIC

PROLOGUE

The plan, which had already elbowed out the other six cycling through Eric Leuthardt's head that day, was to peel back the scalp, pry open the skull, and install a web of electrodes atop D. Brookman's brain. Barring any unforeseen circumstances, Leuthardt hoped to slip an additional phalanx of sensors behind Brookman's left hemisphere—but not before he'd plunged five more deep into the brain itself, tunneling through its mille-feuille of neurons to home in on the epileptic root. A routine procedure, to be sure, and if all went as planned, Leuthardt would have the scalp stapled shut by noon, allowing him time to update the family before heading into an ideas session about enzyme-based treatments for Alzheimer's disease.

It was a typical morning for Leuthardt, rising before dawn for surgery and then trekking from the operating room to the research bench, an invention session, or a meeting with medical device manufacturers. In his dealings with others, he had proven to be collaborative, intense, and eager to please. Now in his early forties, Leuthardt had a muscular build, a thick head of brown hair, and a sloping nose that supported a small pair of glasses. He also possessed that fabulous ability to yoke together worlds previously discrete, binding brain surgery to research, research to the private sector.

Today would be no different. Leuthardt hoped his electrodes would act as a sort of neural seismometer, helping him to pinpoint the source of Brookman's seizures. But diagnostics were only part of the project. Leuthardt's electrodes would serve a second function as well: by melding his platinum sensors to the brain, he hoped to eavesdrop on Brookman's thoughts, creating a link between his patient's brain and the digital world outside.

Leuthardt wasn't angling for thoughts as we know them—conscious considerations like "I feel hungry" and "what a lovely sunset." He was after the electric current of thought itself: the millions of electrical impulses, known as action potentials, that continuously volley between the brain's estimated 100 billion neurons. Those neurons are connected by an estimated 100 trillion synapses, the slender electrochemical bridges that enable the cranium's minute universe of cells to communicate with one another. Like an exponentially complicated form of Morse code, the cells of the brain exchange millions of action potentials at any moment, an electric language that physically underlies our every movement, thought, and sensation. These are not sentient thoughts, per se, but in sum this mysterious and crackling neural language is what makes consciousness possible—a sort of quantum programming code that remains all but unrecognizable to the consciousness it creates.

Leuthardt's hope was to understand that language. Using electrodes to ferry Brookman's neural signals into a nearby computer, he would forge what's known as a brain-computer interface—a wildly intricate union of synapses and silicon that would grant his patient mental control over computers and machines. As this pulsing language streamed from Brookman's brain, the machine's algorithms would work to find repeated patterns of cellular activity. Each time Brookman would think, say, of lifting his left index finger, the neurons associated with that action would crackle to life in a consistent configuration. Working in real time, the computer would analyze those patterns, correlating them with specific commands—anything from re-creating the lifted finger in a robot hand to moving a cursor across a monitor or playing a video game. The end command hardly mattered: once Leuthardt's computers had adequately decoded Brookman's neural patterns—his thoughts—Leuthardt could conceivably link them to countless digital environments, granting Brookman mental control over everything from robotic appendages to Internet browsers.

It's a union whose potential beggars the imagination: an unprecedented evolutionary step—effectively digitizing the body's nervous system—that conjures images of not only mental access to everyday objects like computer networks, appliances, or the so-called Internet of things but also telekinetic communication between people and cyborg networks connected by the fundamental language of neural code.

Just as the body's nervous system comprises both sensory and motor neurons, the wired brain offers an analogous two-way means of communication. Brookman's brain-computer interface may give him control over computers, but it would also grant Leuthardt's computers access to Brookman's brain—a powerful research tool to study the behavior of individual neurons as well as deliver new forms of sensory information.

"We may actually for the first time be able to interact with the world in a nonmuscular manner," Leuthardt said. "I've always needed muscle to communicate with you by moving my vocal cords or giving a hand expression or writing a note or painting a painting—anything. But that may not be the case anymore. So how does that change us? You unlock the mind and make it accessible to science and technology, and suddenly all this other stuff becomes possible. Everything changes. It's a whole new palette for the human imagination."

•

By the time he arrived for pre-op at the Barnes-Jewish Hospital complex in St. Louis, Brookman had already slipped into an anesthetic stupor. "Waaaaassup," he murmured through an opiate haze as nurses wheeled him into Neuropod 5, a surgical unit deep within the hospital. Outside, the September morning was cool and crisp. But here in the sterile confines of the operating room, nature had been all but banished as Leuthardt prepared to remove a section of his skull.

At thirty years old, Brookman had the lean, taut-skinned look of an athlete. His cheekbones were high and wide. They sloped sharply toward his delicate wedge of a chin. His hazel eyes were set far apart, and he trimmed his brown hair short—but not so short as to reveal the surgical scar that snaked around his head like the seam of a baseball.

That scar, the shiny memento of a failed surgery three years earlier, was on full display as nurses hoisted him from the gurney to the operating table. A network of washer-like rings studded his shaved scalp, and his bare chest, draped to the armpits in a blue cotton sheet, shuddered with each shallow breath. His hands lay listlessly by his sides, while his right eye traced blind arcs across the room.

Brookman had come to St. Louis in a last-ditch effort to vanquish the rare form of epilepsy that had gripped him since infancy. Nearly half of all epileptic seizures originate in the temporal lobes, twin brain structures that wrap around the side of the brain like a pair of folded wings. But Brookman's seizures erupted in the parietal lobe, a mysterious brain region that helps unify our experience of touch, sound, and other sensory information.

Seizures in the parietal lobe will often cause pain to radiate from a person's head to his arms and legs. Occasionally, they cause hallucinations. Some people lose their ability to process language or will perceive their bodies as wildly contorted. In rare cases, they orgasm.

Brookman, on the other hand, had none of these symptoms. Though his seizures originated in his parietal lobe, they moved like sheet lightning across the rest of his brain—engulfing multiple lobes and rendering him unconscious. As an adolescent, he suffered seizures during sleepovers or at school. His condition worsened as he matured, and by the time Brookman entered high school, he would regularly fall stiff-limbed to the floor at parties and in classrooms. He lost consciousness on the baseball diamond in front of teammates and at home before his family. He later gave college a shot, but as he put it, "You overwork your brain, and you'll fall into a seizure."

Now thirty, Brookman was out of work. He lived far from his family in a rented apartment and relied on government assistance for his daily drug regimen—a 1,700 mg pharmacopoeia that overflowed with names like Lyrica, Lamotrigine, and Carbamazepine. Nevertheless, Brookman's seizures continued to visit him nightly, often twice, when he would awake parched, sweating, and alone on his bedroom floor. "The medicines work," Brookman's neurologist, Edward Hogan, said. "But he is part of a subgroup of people that no matter what you do, the seizures just won't go away."

Riding just behind the brain's frontal lobe, the parietal lobe plays a critical role in math and in written language. It is associated with synthesizing various sensory inputs, and it is a thoroughfare for vision, whose neural pathways pass through the region as they extend from the optic nerve to the visual cortex in the rear of the brain.

But if parietal-lobe-based epilepsy is rare, the surgery to correct it is even more so. The procedure accounts for only 5 percent of all epilepsy surgeries, in part because resections in the parietal lobe, with its rich intersection of critical brain functions, often cause too much damage to merit the removal of the epileptic source. The surgery has a miserable 50 percent success rate, and even the successes often leave patients with significant postoperative "deficits"—trading partial blindness for the elimination (or reduction) of seizures. The surgery can also damage the patient's ability to write, solve simple math problems, or count fingers on a hand—a constellation of symptoms known collectively as Gerstmann syndrome.

Those were the odds Brookman failed to beat in 2008, when during his first surgery Leuthardt removed a golf-ball-sized chunk of brain from his parietal lobe. Brookman didn't have a seizure the day following the operation, and he briefly imagined he'd landed on the successful side of 50. Within forty-eight hours, however, his seizures returned in all their fury, and as Brookman left the hospital, he despaired at a life of inoperable epilepsy, nightly seizures, and heavy medication.

"Why would God do such a thing?" he asked a few days before the second surgery. "It's like there's never going to be an answer for that. I know there's a God, and I believe in him, but why couldn't he create a miracle for me?"

•

"One, two, three!" Leuthardt counted as the team of nurses lifted the sheet, rolling Brookman onto his right side. Working quickly, they wrapped him burrito-like in egg crate foam, locking him in place with cushioned wedges and thick Velcro straps. As the nurses made micro-adjustments to the table—shifting its foot down by an inch, elevating its head by centimeters—Leuthardt, who wanted it tilted another five degrees, nearly ordered a new one when after twenty minutes the table wouldn't obey.

"If it's not right, it's not right," he said, crouching in blue scrubs to remove the table's boxy control panel. "This is like tuning an instrument before you play: if it's out of tune, the whole symphony will sound terrible."

Cranial position. Clarity of access. Surgical trajectory. These were the concerns that streamed through Leuthardt's head as he held the chunky panel in his left hand, jabbing its buttons with his right. He was preparing to carve a sand-dollar-sized disk of bone from Brookman's skull. The head's position was crucial. A misstep here would reverberate throughout the surgery, complicating each step and cracking the window for mishap. "We're getting a new bed if we have to," he said, working the control panel with his thumb. "It sounds mundane, but it takes years to learn how to position the head. If it's not correct, it will affect every single . . . ," he was saying when through some mysterious digital alchemy the table chimed to life, lowering on command. "All righty!" he erupted, lowering the table and returning the control panel to its cradle.

Meanwhile, Thomas Beaumont, a tall and laconic junior resident who was assisting Leuthardt in the morning's surgery, fetched the Mayfield clamp—a medieval-looking vise that attaches at the table's headrest to immobilize the skull with three steel pins. Someone had written "yes" in blue pen just behind Brookman's left ear, and as Leuthardt positioned the head, Beaumont removed the table's cushioned headrest to insert the clamp. Leuthardt cupped the skull. Beaumont advanced the pins. They were lining Brookman up for optimal access, carefully angling his skull's "yes" region so it would be best positioned for Leuthardt's scalpel. Beaumont twisted the Mayfield's knobs. He paused briefly, retracting the pins as Leuthardt made one final adjustment. Then he drove them home, puncturing Brookman's scalp and locking the skull in place with sixty pounds of pressure. Leuthardt checked his resident's handiwork, giving the head a brisk shake, but careful not to disturb the corona of rings he had pasted on the scalp.

With its knobs, pins, and advancing screws, the Mayfield clamp was one of few tools whose purpose seemed physically evident in a room otherwise awash in the modern technologies of brain surgery. A multiarmed boom carried a dual-lensed camera overhead and several domed surgical lights. Twin monitors pressed against one wall, while smaller screens perched on tables and roosted atop portable workstations. The room looked like the bridge of the starship *Enterprise*, but instead of displaying remote nebulae or distant planets, each monitor displayed a triptych of Brookman's brain—a world that is at once as intimate as the self yet as obscure as a distant galaxy. The spectral images revealed Brookman's brain as a silvery, claylike topography of ridges and fissures. The images comprised hundreds of millimeter-thin virtual cross sections, slicing the brain vertically, horizontally, and laterally. They diced Brookman's brain into an orderly matrix of tiny cubes, enabling Leuthardt to roam freely through this remote world and chart his surgical course to the millimeter. Detailed as Leuthardt's maps were, however, one area remained stubbornly obscure: the grayish void where three years earlier he had removed a spherical chunk of tissue. As he scrolled through the virtual sheets of Brookman's brain, the void stared blindly like a foggy Eye of Jupiter. It was at the border of this absence that Brookman's seizures still gathered.

But first Leuthardt had to synchronize the digital map of Brookman's brain to the organ itself. Taking up a reflective wand that communicated with the camera overhead, Leuthardt placed its shaft in the rings that crowned Brookman's head. He probed each ring, jostling the wand slightly until he hit the sweet spot—the precise corresponding location on the digital map. Pressing a foot pedal, Leuthardt pinged each location, marrying Brookman's physical brain to its virtual twin.

It was an astonishing piece of technology, but perhaps no more so than the golf-cart-sized microscope that hulked in one corner of the room or the crochet-needle-like laser Leuthardt sometimes uses to burn brain tumors from the inside out. Meanwhile, the business end of the room—a buffet of low-tech tweezers, forceps, scalpels, and drills—gleamed against a far wall as strains of Regina Spektor played from Leuthardt's iPhone over the sound system.

Removing the rings, Leuthardt used a salmon-colored sponge to bathe Brookman's head with a sterilizing wash. He gave the scalp a quick shave, lathered it again, rinsed it, and then used a blue marker to trace the horseshoe-like scar he'd cut three years earlier—now adding an oxbow for a little more room to navigate.

Within moments, Brookman began to disappear as Leuthardt papered him over with successive layers of surgical dressing. Using a translucent adhesive, Leuthardt demarcated the surgical site, which he again bathed in disinfectant. He bordered the incision area with a blue sheet, attaching it to the scalp with a stapler.

Swaddled in a nest of sterile sheets and deep in his anesthetic slumber, Brookman now seemed remote. The slight rise and fall of his chest was discernible, but otherwise the only indication there was a living patient in the room was the shaven patch of scalp that poked out from beneath the sheets.

Grasping a scalpel, Leuthardt carved an arch-shaped incision that followed Brookman's previous scar. The tool, known as a bipolar, simultaneously cut and cauterized the incision. The scent of singed flesh began to fill the room as rivulets of bright, oxygen-rich blood trickled into a plastic bag below. The surgeons lined both sides of the incision with blue tourniquet clips to stanch the bleeding and then quickly peeled back the flap of scalp to reveal Brookman's skull.

"Is this Cake?" Beaumont asked as a new song came over the speakers.

"Yeah," Leuthardt replied, using a tiny screwdriver to remove the snowflake-shaped titanium plates he had installed after Brookman's first operation. "It's the new album. It's really good."

Using a pedal-operated bone drill, Leuthardt bored a series of holes in Brookman's skull. After switching to a high-pitched bone saw, he connected the dots, loosening the skull flap before dropping it into a plastic baggie with nutrient-rich liquid.

"How's the baby?" Beaumont asked. "How's the wife adjusting?"

Leuthardt started to say that his young family was fine and that babies only need three things. But the mood in the room quickly shifted as the surgeon studied Brookman's dura mater, the sheath of leathery tissue that encapsulates the brain.

Silver and shimmering, Brookman's whorl of exposed brain looked like mercury against the blue surgical sheeting. It had a gruesome beauty, but Leuthardt had more practical concerns. This was Brookman's second round of surgeries. His dura mater was terribly scarred, and the fibrous tissue had melded to the brain "like a panini." It would be difficult to remove it without causing bleeding on the surface of the brain.

His small talk interrupted, Leuthardt worked silently for the next several hours, alternating between a metal probe, a pair of micro-scissors, and the bipolar to peel away the dura, which clung to Brookman's brain like an old sticker. It was slow, delicate, tedious work. The dura would not release from several of the brain's surface veins, and at a certain point Leuthardt silently considered aborting the surgery altogether.

"It's really moment by moment," he later said.

The surgeon eventually managed to peel back the sheath, allowing him to install two electrode grids—clear plastic sheets containing a total of sixty-four sensors—and five depth electrodes, which he injected deep into Brookman's parietal lobe.

When he began the surgery, Leuthardt had hoped to slide several experimental surface electrodes into the deeper folds of Brookman's brain, but the dura was simply too tightly bonded to the brain. He couldn't slip them under without painstakingly removing even more of the protective sheath, so he settled for the larger clinical grids (grids, incidentally, that were of Leuthardt's own patented design). After replacing the skull flap, the surgeon quickly tunneled the dozen or so electrode wires through one of the boreholes. He trained the wires under the scalp and pierced an exit for them through the skin.

As with Brookman's earlier surgery, today's procedure had been the diagnostic first act of a two-part operation. Brookman's swelling brain would seal the electrodes tight against the neural tissue, enabling neurologists to monitor his seizures over the next few days. After using the electrodes to triangulate the epileptic root, Leuthardt would perform a second surgery, opening Brookman's skull one last time to remove the source of his seizures.

In the meantime, though, Brookman would become a sort of temporary cyborg as Leuthardt's electrodes enabled him to connect mentally to the digital world beyond.

1. BYPASSING THE BODY

Like generations of neurosurgeons before him, Leuthardt had implanted the clinical grid of electrodes to measure the brain's action potentials, tiny pulses of electricity neurons emit each time they exchange information with nearby cells. The technique, known as electrocorticography, or ECoG, doesn't record individual neurons. Rather, the grid's electrodes pick up the collective activity of the thousands of neurons that lie beneath them, registering their summed rhythms as brain waves.

By tracking Brookman's brain waves, neurologists could observe when his normal brain activity was interrupted by the beginnings of a seizure. Instead of the normal up-and-down signal of, say, an alpha wave, Brookman's brain would become erratic as a cluster of neurons began firing in synchronic bursts. The renegade cells would inevitably recruit more neurons to their ictal cause, triggering Brookman's brain waves to grow chaotic as the epileptic storm pulsed across the brain.

It had been crucial during the implantation surgery for Leuthardt to install sensors over the entire seizure focus area. Ample coverage would enable neurologists to divine an epileptic source by noting, essentially, that brain waves first became erratic below a specific electrode. "It's kind of like a murder mystery," Leuthardt said. "You're trying to find the criminal. You can use external studies like MRI and PET scans. Those will tell you the general region—the criminal's zip code. But now we need to find his address. Electrodes give us very specific localization."

Using a similar technology, neuroscientists have long listened in on individual neurons with penetrating electrodes. Piercing their hair-thin wires into the brains of monkeys and rats, these scientists spent years searching for repeated firing patterns in individual cells. As the animals performed repetitive gestures like pressing a lever for a juice reward, the researchers found that specific neural patterns were associated with the physical action. There was a lot of room for error, but the neural patterns were often consistent. They were also repeatable: a neuron would erupt in a similar firing pattern each time the animal performed the bar-pushing action to receive its juice reward.

Around 2000, however, a handful of researchers began transforming this information into a brain-computer interface: whenever the cell produced the desired firing pattern, the computer would execute a physical command. Of course, most animals were none the wiser and would continue to press the lever to receive their juice reward. But over time, they realized they didn't need to physically press the lever to get their reward. They had only to *think* about it.

The clinical application for brain-computer interfaces seemed clear. Physical paralysis essentially is a communication error between the central nervous system and the branching network of peripheral nerves that radiates from the spine. In healthy bodies, the brain sends signals to the spine (*Walk! Sit down!*), which in turn relays the details to the peripheral nervous system (*Lift the right leg! Bend at the hip!*). Paralysis occurs when some relay point along the route stops working—be it through spinal cord injury, amputation, stroke, or some other form of neuron death. The brain may continue to send signals, but the message never arrives.

By the time Leuthardt entered the field in the middle of the first decade of the twenty-first century, researchers had already shown that by sinking individual electrodes into the brains of monkeys, rats, and some humans, they could tap movement at its source. A single neuron provided enough information to create basic computer commands, bypassing an animal's peripheral nervous system to give subjects modest neural control over machines. In some cases, this meant linking neural patterns associated with moving a joystick to give monkeys direct control over a cursor. In others, it meant controlling a robotic arm or mentally pressing a lever to deliver a juice reward. The control was basic, enabling the animals to move a cursor to the left or the right, or a feeding bar up or down.

But these early researchers were tapping only a handful of neurons. What sort of control could they produce if they harnessed, say, a hundred or even a thousand neurons in a person? Could they give people full control over a computer, enabling them to send e-mail or surf the Internet using only their thoughts? Could they re-create the elegant movements of the human arm? Harnessing thousands of neurons, could researchers craft a full-body exoskeleton for quadriplegics or soldiers? And how about abstract thoughts? Given ample neural access, could we bypass spoken language altogether, doing away with its ambiguities and miscommunications in favor of direct neural exchange? In the realm of memory, could brain-computer interfaces enable total recall? Could they deliver new sensory modes like infrared or X-ray vision? What was to stop these technologies from enhancing our own cognition? Could we selectively stimulate the brain to boost learning?

Those early brain-computer interfaces might have been confined to basic physical commands, but Leuthardt saw in them a union that could fundamentally change our understanding of the brain. "I saw neuroprosthetics in the very early, seminal stages," he said, "and I thought, this is it. This is the future."

Leuthardt was not alone. The field was already thick with speculation that scientists could craft a neural augment for people with paralysis. In 1998, an Irish researcher named Philip Kennedy demonstrated that he could endow a man paralyzed from the neck down with rudimentary control of a computer program. One year later, the German researcher Niels Birbaumer used EEG to enable similarly impaired patients to control basic word-processing software, and by 2001 one of the field's titans, a neuroscientist named John Donoghue, cofounded Cyberkinetics, a neurotechnology company aimed at developing commercial brain-computer interfaces. Other researchers were using electrodes to unlock the brains of monkeys. In one headline-grabbing experiment, Duke University's Miguel Nicolelis connected the motor cortex of a rhesus monkey to a robot arm in the next room. Using only its thoughts, the animal harnessed the arm to play a simple video game. "At that moment," Nicolelis wrote, "the cumulative years of research and the hopes of thousands of severely paralyzed people who dreamed of one day regaining some degree of their former mobility became deeply intertwined."

Still, there was a lot of work to do. These early efforts were a far cry from the sort of always-on commercial device Leuthardt envisioned. And that's to say nothing of crafting a brain-computer interface, or BCI, to rival the elegance and diversity of biological movement.

What's more, the interface itself was problematic. Penetrating electrodes might have enabled brain researchers to enter an intimate exchange with the brain's most basic unit—the neuron—but they were also unreliable. Like the rest of the body, the brain abhors foreign objects, and while the platinum sensors created a close union between mind and machine, it was often short-lived. The brain eventually mounted an immune response, dispatching micro-glia, astrocytes, and other proteins to cordon off the offending electrodes. Wrapped in successive layers of scar tissue, the electrodes inevitably lost their sensitivity. Signal quality degraded, sometimes in a matter of months, rendering the implant unusable. "There was no way that was going to work," Leuthardt thought. "If these microelectrodes were not lasting longer than six or seven months, there was no way a neurosurgeon would ever want to put this into a patient commercially."

Electroencephalography, or EEG, was an option, but surface electrodes had their own problems. It was a rare individual who would be willing to spend his life in what amounts to a sensor-studded swimming cap. More important, though, surface electrodes provided only a hazy portrait of the electrical storm raging inside the skull. Placed directly on the scalp, EEG electrodes can't always differentiate between the electricity inside the brain and the electrical pulses that animate the scalp. It leads to a muddy signal, adulterated with muscular electricity and even surrounding electronics.

At the time, researchers confined themselves to either EEG or penetrating electrodes. Those interfaces were fine for the research lab, but Leuthardt was convinced that if he and his fellow scientists were ever to usher in the age of neuroprostheses, they would need to enter the commercial market, crafting a highly sensitive, accurate interface that wouldn't degrade over time.

"That's what got me down the road of ECoG," he said. Unlike penetrating electrodes, the ECoG grids did not pierce the brain. Rather, they rested on its surface and would likely be more stable. Having direct contact with the brain also meant that, unlike EEG, ECoG signals weren't as likely to be contaminated by muscular artifacts from the scalp or nearby electronics. It seemed like the Goldilocks zone: more stable than penetrating electrodes, more precise than EEG. "I've always seen us as being the bed's just right in the sense that this one is too invasive, that one is too noisy, but this one is just right."

If an EEG was like listening to the muffled strains of the neural symphony behind a band shell, penetrating electrodes were like training a microphone on a sole musician or an individual string. ECoG, by contrast, was like listening to a section of the orchestral brain from the first few rows—close enough to tease out the first violins from the second violins.

But here was the real beauty of using ECoG: as a neurosurgeon, Leuthardt already had a built-in population of human research subjects. During the week or so that patients like Brookman were implanted for epilepsy monitoring, they were effectively lying in a hospital bed just waiting to have seizures. The rest of the time? The electrodes simply sat atop the brain, passively recording its electric hum. All the elements were there. Why not use the clinical setup of the epilepsy-monitoring unit to create an entirely new brain-computer interface?

At the time, all but a few neural implants were used for limited periods of time and only in the laboratory. But a neural implant that could pull detailed information from the brain while also sidestepping the glaze of signal-degrading scar tissue? A device like that could form the basis of a commercial implant that would remain in the brain for years. "It became very clear to me that this was the future," said Leuthardt. "It's a whole new universe that opens up—one that can change the human experience."

•

To that end, David Bundy arrived at the epilepsy-monitoring unit a few days after Brookman's surgery with a cartload of electronics. As a graduate student in Leuthardt's research lab, Bundy was hoping Brookman would don a sensor-studded glove. He wanted him to flex his fingers so he could calibrate the movement to Brookman's brain activity, the first step in building a BCI.

But Brookman was still recovering from surgery. His eyes fluttered and his head nodded lazily as he slouched semiconscious in the hospital bed. Shirtless, he wore a pair of thin cotton shorts, and his head was wrapped in a turban of gauze dressing, a Gorgon-like mane of wires exiting the right of his skull.

Normally, the tangle of wires that spilled from his head would transmit Brookman's brain waves to a bank of computers down the hall. But Brookman had agreed to be one of Leuthardt's research subjects, and for an hour each day grad students like Bundy connected his cables to their own cart of amplifiers, digitizers, and computers.

Leuthardt kept the amplifier in what's known as a Faraday cage, a wood-framed box wrapped in copper mesh to isolate the device from surrounding electronics. From one side of the cage tumbled a rainbow-colored cascade of wires that linked the amplifier with the leads exiting Brookman's brain. From the other, the amplifier connected to a computer whose screen showed a graph of brain waves from each electrode.

It was a surprisingly ad hoc affair, with single-serving cups of orange juice, travel-sized bottles of Listerine, and toothbrushes still in their wrapping strewn across the room. These signs of the family's vigil were everywhere: a half-eaten cluster of grapes sat on a table next to individual-sized bottles of body wash and shampoo.

Meanwhile, Brookman's mother and aunt watched warily from a pair of vinyl-covered lounge chairs. The room's blinds had been drawn against the morning sun, and a nurse in maroon scrubs sat quietly in the corner, prepared to intervene should Brookman seize during the testing.

It was no empty measure. One day earlier, Brookman's eyes had rolled back in his head and his body stiffened just as the grad students were setting up their equipment. They beat a quick path to the door as Brookman fell into convulsions, the day's research session scuttled.

Now Brookman seemed only slightly awake. "We want to see what your brain signals are doing when you're moving your hand in different ways," said Bundy, explaining how they wanted to correlate the movements of the sensor-laden glove to specific brain waves. "The goal is to help people that maybe have spinal cord injury or amputation so they can have a prosthetic hand."

Bundy told Brookman he'd be following a series of simple prompts to link, or calibrate, his brain waves to the movement of his hand. He'd need to flex his thumb, extend his index finger, and pinch with his thumb and index finger. Once they'd calibrated the glove, they would move on to the task itself: Brookman would think about making a specific hand gesture to mentally control the up-and-down movement of a column on the monitor.

"Does that sound all right?" Bundy asked after explaining the task.

"Yeah," drawled Brookman, only half-awake.

"Does that make sense?"

"Yeeeeaaaah."

But it didn't make sense. Brookman lagged behind the simple prompts, incorrectly pinching or extending his forefinger five seconds after the computer prompt. By then, the computer had moved on to the next prompt, and Bundy had to start the program again after a few failed tries.

"Just flex your thumb and then extend it out," Bundy said. "A pinch would be just bringing your thumb and your finger together—just like this," he said, making an "okay" sign with his right hand.

"Just your thumb, baby," Brookman's aunt interjected. "Keep your hand open and just do your thumb. Are you awake, baby?"

Brookman was, but only barely. Though he normally took a host of antiseizure drugs, neurologists had taken him off his medication to better locate his seizure focus. "We want to make sure we get the seizures, because occasionally we'll put all these electrodes on, and they won't have any seizures," said Brookman's neurologist, Hogan. "We were pretty aggressive."

The neurologist needn't have worried: without medication, Brookman had suffered some twenty-five seizures in the first twenty-four hours following the surgery, more than Hogan had ever seen. Emerging disoriented from these rolling convulsions, Brookman didn't know where he was or what had happened. He'd been wild in his panic, trying to rip the wires from his head and lashing out. It got so bad that at one point the hospital staff restrained him with leather straps.

But now Brookman was dazed and docile with pain relievers. He was meek, eager to work with the researchers, and fearful he would disappoint them.

"Can you understand what he's saying?" his aunt asked.

"Yeeeeaaaah," Brookman moaned as the computer prompted him to make a fist.

"Can you make a fist?" she coaxed as he brought his fingertips slowly to his palm. "Good job!"

"Can you flex your thumb?" Bundy jumped in, following the computer prompt.

"Just your thumb, baby," said his aunt, a woman with spiky brown hair and a peach-colored blouse. "Do it with your thumb."

But Brookman moved both his index finger and his thumb, moving them slowly in unison.

He clearly wasn't up to the task. Brookman's seizures, coupled with the pain medication, kept him semiconscious. He was easily confused, nodding off in the middle of tasks and unable to follow the simple instructions.

"I think we want to just let you rest," Bundy finally said after several failed attempts. "We might try to come back in a little bit."

"I'm ripped up," Brookman apologized.

"We'll let you rest."

"Let me rest to where I can at least see straight," Brookman said. "I'm so tore up right now."

"That's understandable," Bundy responded.

"No matter what, I promise to God and cross my heart I'll make sure I get the job done right," Brookman said. "I'll make sure they get the best possible stuff."

"It's okay," his aunt said. "They know you will."

But it was too late. Brookman was becoming upset, his eyes brimming with tears and his drug-lazy voice tensing with frustration.

"I just can't see straight," he said.

Bundy, a Texan with a full beard and large ears, shifted near the bed, made uneasy by Brookman's frank emotion. Meanwhile, another grad student, Nick Szrama, ventured that Brookman was already "helping out quite a few people" as he and Bundy began packing up their research equipment.

"Okkkaaay," Brookman murmured. "If I could see straight, I'd be able to do this."

•

Challenging though they are, difficult research conditions are in some ways the least of Leuthardt's concerns. The neural matrix is wildly complex. We understand very little of even the brain's most basic functioning, and its three pounds of neural tissue do not readily yield their secrets to the system of 1s and 0s Leuthardt and his cohorts would use to reveal its mysteries. And that's to say nothing of the more basic biological problem researchers encounter when they try to join the hard stuff of electrodes to the squishy tissue of the brain.

With so many unknowns, Leuthardt's vision of creating a meaningful union between mind and machine could ultimately remain little more than a twenty-first-century parlor trick—clunky and limited, but catnip to futurist nerds whose imaginations catch fire each time a researcher with an electrode cap crops up on YouTube. His dream derailed, Leuthardt may one day be remembered only as a neurosurgeon who in the early twenty-first century began amassing a superhuman arsenal of intellectual property. At last count, he had more than 860 patents on file. ("Thomas Edison had 1,093," he quipped. "So that's my goal.")

In this telling, Leuthardt's *Wikipedia* page may someday mention that he was born to immigrant parents. That his father moved back to Germany. That he was raised lonesome in working-class Cincinnati by a single mother. That he once published a science fiction novel, dabbled as an abstract painter, and had a yen for objectivist philosophy, futurism, and handguns.

Still, these are but the ornaments of a life, personal statistics that are never all that illuminating. In the meantime, however, Leuthardt couldn't help it: his mind seemed always to be reaching for some essential through line that would create a new opportunity from a current task. He had found just such a line when he forged his research lab from his surgical practice. Those twin enterprises spawned new inventions—brain retractors, electrode grids, novel brain catheters—that inevitably led to new inventions, new patents, and new start-up companies. It was all of a piece for Leuthardt—exponential results, he called it.

Leuthardt's real genius, though, was his knack for temporarily lashing together the minds of academics and clinicians, pushing them to engage their brains in ways that don't come naturally to academics or clinicians. He pressed them into service not to idly toss around a few abstractions while stroking the collective beard. His quarry was something more tangible. He was looking to extract results—technical fixes to problems the rest of us hardly knew existed. That, and to stockpile a defensible armory of intellectual property. "You've got to identify the problem, then you can find a solution, or you find a solution and pair it with a problem that matches," he said. "You don't have to know everything. You engage people who know more than you, and then you create an environment that can accomplish things that none of you could have done by yourself."

Ideas for Leuthardt were not some delicate species that crept quietly in the night. Nor were they violent strokes of insight that flashed through the mind of the toiling genius. They certainly didn't come out of thin air. For Leuthardt, ideas were like reptilian young. "You spawn a lot of them," he said, "and see which ones survive." The trick was to uncover and nurture them. Ideas percolated during conversations. They teemed forth in the operating room. They sprang from inefficiencies in patient care and emerged after months of painstaking research. But the best ideas occurred at the margins—that intersection, say, between neuroscience, biomechanical engineering, and cardiology—liminal spaces where intellectual outsiders could tackle long-standing problems. He pushed others as he pushed himself. Those he pushed came to believe in the process, their world revealing itself as a series of solvable engineering problems and legally defensible solutions.

"We get criticized for always looking at the possibilities and not being realistic. But nothing good happens if you just focus on what's going to prevent you from getting to the next stage," he said. Problem solving and positive thinking were skills. You had only to train your mind. "You pester your subconscious by constantly trying to think of a solution and not coming up with one. Then you let it go," he said. "You let your internal—that area below conscious awareness—work on it, and invariably something pops up."

This was the sunny, future-tense world Leuthardt was forever saying he liked to escape in the operating room. He insisted surgeries were the most relaxing part of his week, when the yeoman tasks of cutting and sawing and suctioning trumped his business plans and inventions, his collaborations and research. He swore he found it relaxing, meditative even, as the rest of the world receded behind the glare of surgical lights and he could steep himself in the minutiae of the moment, cutting through layers of flesh and bone, excavating the brain. There were no distractions in the OR. There was a purity of purpose where he would often work uninterrupted for eight- or even twelve-hour stretches at a time, a much-needed reprieve from his chase for answers.

Or at least that was the idea.

But with his frontal lobe engaged by the delicate task of slicing through neurons and the not-too-delicate task of sawing through skull, his brain's hindquarters would inevitably begin to sift a problem, subconsciously deconstructing it until, pop! A solution sprang forth.

Leuthardt needed that pop! of the new as much as a marathoner needs a runner's high. A "professional anorexic" was what he called himself, and for all his avowals that surgery was a sanctum and that he longed for its singularity of focus, the future still beckoned.

And nowhere was that call louder than in epilepsy surgery, a two-step operation that not only allowed him to plug his electrodes into the human brain but enabled him to do so for weeks at a time.

•

Leuthardt's idea, or at least his germ of an idea, was to build a company around the core technology of ECoG, making neuroprostheses for the consumer medical market. Once he'd established a beachhead, showing that neuroprostheses were both safe and effective, he believed the technology would spread to other medical uses. "You'll start to see the collateralization of that technology—to spinal cord injury and hopefully traumatic amputation," he said. He was convinced that once neuroprosthetics had successfully established a footing in the medical world, they would eventually achieve something even more momentous: brain-computer interfaces that augmented human ability. "The big leap happens once we become good enough that the implant gives you some type of social advantage. It's all simple stuff right now. We're all playing *Pong*. But *Pong* evolved. *Pong* evolved into Xbox 360," he said. "It's a natural extension of human behavior. If you can change yourself so you can facilitate things you want to do? People will do that. That's the grand horizon. Essentially, you've unleashed the brain on the world."

Today's neuroprostheses may be in beta form, but that hasn't stopped the army from funding Leuthardt and his colleagues' research into language. Working with researchers in New York, the group is trying to decode the neural basis of language, raising the possibility that someday soldiers will be able to communicate using only their thoughts—a sort of digital telepathy. Meanwhile, the Defense Advanced Research Projects Agency, or DARPA, the blue-sky research arm of the Department of Defense, is funding scientists in Southern California who are trying to craft a neuroprosthetic for memory. Led by Theodore Berger, these researchers are working in mouse and monkey models to develop a BCI that would bypass the hippocampus, a sea-horse-shaped brain structure essential to memory. By analyzing the change in neural firing rates as they enter and exit the hippocampus, Berger and his colleagues have developed what they believe is a sort of meta-algorithm of memory. Scientists first disable the animals' hippocampi, ensuring they have no working memory. Then, using electrodes, the researchers record incoming sensory data to the disabled hippocampi in a bank of computers, which processes the action potentials to mimic the function of the hippocampus. Researchers then stimulate the subjects' brains with the transformed firing patterns, creating basic memories with the prosthetic, such as where to find a food reward.

In other research, human subjects can use their brains to control digital avatars, and quadriplegics are again feeding themselves using thought-directed prosthetic limbs.

As with the Cold War push to develop atomic weapons, or the midcentury race to discover the structure of DNA, the government is funding much of today's BCI research. That investment grew in 2013, when President Obama announced the BRAIN Initiative, which adds another $100 million to brain and BCI research. Like those earlier races, the emerging field of neuroprosthetics is filled with warring factions. The competitors are again ambitious and highly accomplished—colleagues turned rivals who compete for government grants, scientific dominance, and fame. "These guys know that there will be a Nobel Prize," said one of the field's giants, Miguel Nicolelis. "It's become really, really competitive."

But unlike those earlier contests, where scientists worked under the banner of a government or university, BCI is coming of age when universities are looking for any competitive edge their employees' intellectual property may bring. Lured by the potentially mammoth payouts of the private sector, scientists like Leuthardt are bucking against the sober confines of traditional academic research. It's an entrepreneurial world, where students are schooled in the art of presenting their findings to medical device makers and researchers are mingling with venture capitalists in the hope of monetizing their results.

Like Leuthardt, many researchers are working to develop BCI for clinical applications, but many are equally, if not more, excited about the technology's potential to amplify human ability. "We don't know how far we can go," Kevin Warwick, a cybernetics researcher at the University of Reading in England, said. "What can we do if we link a human brain more closely to a computer network? What opportunities does that open up? You're into the matrix, and to say, 'Oh no, that's just science fiction . . .' Well, no."

Warwick made history in 2002 when he had a grid of a hundred microelectrodes implanted in the medial nerve of his left arm. It wasn't a direct cortical implant, but the device picked up neural activity from his peripheral nervous system, enabling him not only to control a robotic hand linked to the nerves in his arm but also to perceive sensory stimulation from the electrodes. As the robot hand gripped an object more tightly, the electrical pulses to the stimulating electrodes increased in frequency. "The brain makes the best sense it can from the signals," he said. "It wasn't likening it to anything else. It didn't think of it as being my hand gripping in terms of my biological hand. It took it on board and used the signals for what they were."

In a final flourish, Warwick's wife, Irena, received a similar implant, enabling the pair to "link" their nervous systems over the Internet. The technology was crude: Warwick's electrodes delivered an electrical pulse each time Irena moved her hand. Rudimentary as it was, however, they had merged their nervous systems in some small way, projecting their movements far into the digital realm to endow the couple with a neural awareness of each other's movements.

For Warwick, limiting the use of a BCI to an exoskeleton or a neurally controlled prosthetic is a "conservative" view of the body. "You're just making the body a little bit more powerful, or giving somebody a slightly more powerful arm," he said. "You can have a completely different concept of the body your brain is controlling. It doesn't have to be arms and legs. It can be any type of technology you want. The whole concept of the body can and will be considerably different." To researchers like Warwick and Leuthardt, neuroprostheses do not only challenge our traditional notions of the human body. Rather, they believe BCIs will fundamentally transform our understanding of the brain, consciousness, and what it means to be human. "If you link your brain to a computer brain with different sensory inputs and different mathematical abilities, you're into this sort of thing where a computer can deal in multidimensional processing," Warwick said. "Instead of thinking as your human brain does in three dimensions, you can start thinking, potentially, in twenty or thirty dimensions. What does that mean? No idea! You're into a whole different world really."

Just this sort of research is already taking place at Nicolelis's lab at Duke University, where researchers have used infrared sensors and stimulating electrodes to enable research animals to perceive the infrared portion of the light spectrum—a "sixth sense," as Nicolelis calls it. In a separate experiment, Nicolelis is using computers to merge the brains of lab animals. Using electrodes to record neural activity from one animal, the scientist uses those same firing patterns to stimulate the brain of a second animal—enabling the second animal to share the experiences of the first. "The brain is so plastic that it can incorporate another body as its source of information to probe the world," Nicolelis said. "If we take this idea really seriously, we could assimilate anything that gets in contact with the brain—including another being, including the body of someone else. That touches on theories of self, theories of identity."

Bolstered by futurist writers like Ramez Naam and his promise that future brain implants will seem "as natural as breathing," these researchers point to Moore's law, the Silicon Valley adage that computing power doubles every two years, allowing devices to become smaller and faster and cheaper. This principle of exponential growth, so named for the Intel cofounder Gordon Moore, has held true since the birth of computing more than half a century ago. It has certainly held true for today's smartphones, which bear little resemblance to the room-sized computers of the 1960s.

Like those early machines, many of today's most advanced brain-computer interfaces are wired and bulky. They often require a cart of computers the size of a dishwasher to function properly, and that's to say nothing of the two or three technicians who must be on hand to keep the interface chugging along.

Nevertheless, the field is already shrinking and enhancing its technologies by creating lighter, more efficient power sources and better neural interfaces that can communicate wirelessly with networks. "When I started twenty years ago, you had to have a roomful of equipment to record twenty neurons," Nicolelis said. "We are getting close to a thousand neurons now, and it's about two inches by two inches, the little chip. It's moving much faster than we expected."

Neural stimulators, implants that deliver small pulses of electricity to specific brain regions or parts of the peripheral nervous system, have been on the market for years. More than 300,000 deaf patients have received cochlear implants, which stimulate the acoustic nerve to approximate natural hearing, since the devices first gained FDA approval in 1984. Similarly, deep-brain stimulation, or DBS, which uses surgically implanted electrode leads to deliver pulses of electricity deep inside the brain, has been approved to mitigate the effects of several movement disorders, including essential tremor and Parkinson's disease. Research studies have shown that DBS can also be effective in treating severe depression and chronic pain. Meanwhile, NeuroPace, a neurotech firm in Washington State, recently won FDA approval for its brain implant that uses small doses of electricity to disrupt epileptic seizures as they emerge. Similarly, the FDA recently approved the Argus II, a visual prosthetic that uses a camera mounted on a pair of glasses and a retinal implant to endow otherwise blind users with a rough approximation of vision.

These early BCIs remain in the medical realm and relatively crude. The Argus II, for instance, contains only sixty stimulating electrodes, something like the equivalent of downscaling a 1080p HD screen to 60 pixels. Although the system's camera may record the entire scene, the image converter must drastically reduce the image to conform to the implant's parameters. The result is a black-and-white image that features mainly objects with definitive lines in sharp relief, like street curbs or doorways.

Nevertheless, the Argus II has helped otherwise blind individuals navigate city streets, and the company is at work on future models with color vision and even zoom lenses—innovations that lead some futurists to foresee a day when neuroprosthetics will enhance human ability.

"As you lie there on the operating table, the doctor makes a tiny hole in your skull, through which she inserts an incredibly light, flexible mesh of electronic circuits," Naam writes in his transhumanist manifesto, *More Than Human*, imagining a day when elective neural implants are as common as teeth whiteners are today. Over time, Naam imagines, "you routinely trade memories and experiences with other implanted humans. You learn to view the world through other people's eyes. You let others see through yours. As the months and years pass, you increasingly view your implant as a vital and natural part of you. Using it becomes as natural as breathing. You can no longer imagine a disconnected life."

Undoubtedly, years of research and a thicket of scientific and technical hurdles must be cleared before people will start lining up for elective wireless implants that can be synced to HVAC systems or the Internet. But while researchers like Leuthardt recognize these many obstacles, they remain convinced that this future is on the not-so-distant horizon. "Science progresses accidentally and sometimes exponentially. Very rarely do you get a linear progression," said Leuthardt, who believes elective neural implants will be available inside of two decades. "So is it unreasonable to think about people using these things to enhance their abilities? If these things are minimally invasive or noninvasive, easy to apply, and easy to use? Probably not."

•

In some essential sense, we've been enmeshing our lives with tools ever since *Homo sapiens* emerged from the hominid line some 200,000 years ago. Be it a spear, fire, eyeglasses, a computer, the printing press, or the wheel, one trait that separates humans from most other animals is our sophisticated ability to fashion diverse technologies to amplify our power, intelligence, and abilities. These tools quite literally become us. Many remain outside our bodies, but with others (like pacemakers) we enter a relationship so intimate that the nonbiological device disappears as it is integrated into our own self-perception. The tool becomes invisible, an augment that, while not inborn, we nevertheless adopt as our own. Medical devices like pacemakers and hip replacements are only the most obvious examples of assimilated technologies, but tools that are not physically merged with the body can also become so integral to our consciousness that they are all but invisible.

Take writing, a technology in that it requires tools and is not an innate trait. Writing is a skill. It must be acquired. Plato himself was skeptical of the transformational power of writing. In *Phaedrus*, the philosopher wrestles with the primacy of speech, which he considered natural, versus writing, which he deemed a shabby counterfeit. "There is something yet to be said of propriety and impropriety of writing," Socrates tells Phaedrus before recounting the story of the Egyptian god Theuth, whom he credits with inventing the "use of letters." Socrates recounts how Theuth presented his invention to Thamus, a greater god. Theuth argued that writing would make Egyptians "wiser and give them better memories." But Thamus was skeptical. He worried that writing would cause people to become lazy and stop using their memories. "They will be hearers of many things," Thamus said, "and will have learned nothing."

Of couse, Socrates famously never wrote anything down. He left that to his student Plato, who extended the cultural memory of his Socratic dialogues. In essence, it is only through a medium Socrates loathed that we are able to even approximate what he said about writing.

But writing is more than merely a historical record or a tool to relieve the burden of memorization. Rather, writing is a technology we integrate into the brain that allows us to perform cognitive tasks we would otherwise be unable to achieve. Be it thinking through a complex ethical issue or working out an algebraic equation, writing acts as a sort of external memory device, a cognitive augment that allows us to organize our thoughts and break down complex problems into more manageable steps. As the philosopher Andy Clark observes in *Natural-Born Cyborgs*,

> The brain learns to make the most of its capacity for simple pattern completion (4×4=16, 2×7=14, etc.) by acting in concert with pen and paper, storing the intermediate results outside the brain, then repeating the simple pattern completion process until the larger problem is solved. The brain thus dovetails its operation to the external symbolic resource. The reliable presence of such resources may become so deeply factored in that the biological brain alone is rendered unable to do the larger sums . . . Many of our tools are not just external props and aids, but they are deep and integral parts of the problem-solving systems we now identify as human intelligence. Such tools are best conceived as proper parts of the computational apparatus that constitutes our minds.

It's not merely that writing or the use of symbolic figures to represent numbers enables our brains to process more abstract or complex problems. Rather, these tools become so deeply embedded in our intelligence that they essentially disappear. We identify the cognitive augment not as something outside us but rather as something that defines us.

Writing may be an extreme example of technical integration, but neuroscience is beginning to show that there's some scientific truth to the old adage about tennis players' becoming "one with the racket." In fact, the neuroplastic brain actually undergoes physical changes after repeated use of a tool, expanding its map of the body to include tools like tennis rackets or eyeglasses.

Studying the brains of right-handed violinists, German researchers have found that the sensorimotor area of the brain that corresponds to the left hand (which right-handed violinists use for the instrument's fingerboard) is larger than in nonmusicians. More pointedly, a group of Japanese researchers studying monkeys found that the animals' brains actually regarded certain tools as part of their bodies. Using electrodes to record the animals' neural activity, the researchers first touched the animals' hands and arms to identify how the monkeys represented that area of the body in neural space. As they continued recording, the researchers gave the animals rakes, which the monkeys used to drag pieces of food from behind a screen. The rakes were the animals' sole means of gathering the reward, and researchers allowed the monkeys to use the rakes for several weeks. Once the animals were accustomed to the activity, researchers began to touch the rakes. The results were astonishing: When researchers handled the rakes, the same portion of the monkey's brain that had fired upon feeling its hand touched began to flare up. The monkeys had mentally *embodied* their tools, which their brains represented as an extension of the arms and hands.

From pacemakers and contact lenses, to cars, social media, and hip replacements, the distinction between what is us and what is our technology has never been murkier. But while we have always integrated cultural tools like writing, we are now absorbing technology directly into our bodies, be it through implants or wearable technologies that increasingly mediate our social, personal, and professional lives. "Machines are becoming more and more enmeshed in our personal sense of ourselves," said Leuthardt. "This notion of advancing technology and how we are getting closer and closer to the tools we use, and the notion of body modification—they are all converging to where neuroprosthetics can go beyond being merely a tool for restoring function but actually augmenting function."

Like the brains we house, we are wildly adaptable, and innovations that were once suspiciously regarded as levelers of culture—the outcry over, say, writing—are quickly absorbed into mainstream use. "Look at plastic surgery. Thirty years ago, all of those procedures were for people who had facial injuries or for mastectomies after breast cancer. They were restorative treatments after some distortion of that person's anatomic form," Leuthardt said. "Now there are girls who are eighteen who are getting breast augments before they go off to college. What was once intended to help people with deficits is now a graduation gift."

To that end, Leuthardt and his colleagues founded Neurolutions, a venture-capital-backed company to transition neuroprosthetics from the laboratory to the free market. The company's primary mandate, at least initially, is to restore function in stroke patients. "But then," he said, "the platform becomes available for anything else you want to do with it. The world essentially becomes your iPad."

2. DARPA HARD

Chocolate may seem like a frivolous goal for a Defense Department program—especially one with a budget of more than $75 million that has funded hundreds of researchers. But in a fundamental sense, being able to pick up a bar of chocolate is all that matters—at least for a guy like Geoffrey Ling, the irrepressible program manager who oversees DARPA's Revolutionizing Prosthetics program.

The retired army colonel doesn't so much talk as try to keep verbal pace with the torrent of thoughts that impels him. Words seem like clumsy things for Ling—an outmoded technology that barely manages to convey the rush of ideas, ambition, energy, and passion that drives him. Ling has an easy smile. He brushes his short hair forward and to the right, and he has little patience for scientists who want to pursue science for the sake of science. At least not on his dime. At least not right now. "Anybody who wants to drift, we put him back on the mark," Ling said. "Yes, there's a lot of good science. Feel free to go and do it in your spare time, but right now you've got to do this because you're obligated to do it. That's keeping your eye on the goal. That's how you get there. That's how you make these big advances."

Ling's military background makes him a bit of a rarity at DARPA, but his experience also helped forge his resolve to help wounded soldiers. Ling joined the army after graduating with a medical degree from Georgetown University. He went on to complete a neurology residency at Walter Reed Army Medical Center, later specializing in neuro-intensive care at Johns Hopkins.

But it was his twin tours of duty that convinced him neuroprosthetics could play a critical role in helping veterans who had lost limbs or suffered traumatic brain injury in battle. "There wasn't a day that went by that I wasn't taking care of somebody who had lost a limb," he said, recalling his deployment to Afghanistan in 2003, where he ran an intensive care unit. Most of the amputees were Afghans—kids mainly, who'd run across a discarded land mine left over from the Soviet invasion. "Already you could see the need," he said. "Then, as the war picked up a little bit, we actually took care of some soldiers who had traumatic amputations."

Ling returned home in 2004. He continued working as a neurologist, but in his research lab he was developing a noninvasive imaging technology that caught the attention of Kurt Henry, a DARPA program manager who thought Ling would be a good fit for the agency. "He ran my name up the chain," Ling recalled. "I was tapped. It's like a secret club almost—like Skull & Bones or something."

Ling was soon called to a meeting with Tony Tether, who as the head of DARPA had overseen the Human Assisted Neural Devices, or HAND, program, a multimillion-dollar effort to develop BCIs to "improve warfighter performance on the battlefield" as well as "enhancing the quality of life of paralyzed veterans."

"It was a very DARPA approach," said Ling. "They were really interested in looking at brain signals and seeing how those brain signals were related to actual motor output . . . It was typical of DARPA—a very, very basic science view."

By then, the conflicts in Iraq and Afghanistan were heating up. Advances in body armor—along with the proliferation of so-called improvised explosive devices, or IEDs, the crude bomblets insurgents concealed along roadways and detonated by cell phone—

meant soldiers who might have perished in earlier conflicts were now surviving the attacks, often to devastating effect. Limb amputations were increasingly common, as was traumatic brain injury, which would soon become one of the conflicts' signature wounds.

Within a year of joining DARPA, Ling again went to war. Now stationed in Baghdad, he was running the ICU for the Eighty-Sixth Combat Support Hospital when a national guardsman was rolled in to the ICU. The soldier's Humvee had been caught in an IED ambush earlier that day, blowing him through the roof and breaking his spine in three places. "When I examined him, he was moving everything. He was moving absolutely everything, so his spinal cord was okay, but his bone was fractured," Ling recalled. "I said to him, 'You know, Specialist, you have a million-dollar wound here.' In World War II, that was the wound that was bad enough to get you out of the combat zone, but not so bad that you would have a lasting deficit." Ling told the kid he was going to be fine. He was going to be discharged. He was going to get to go home and get better. He'd have to leave Iraq. He'd have to leave the army, but his back would heal completely; he'd even be able to play sports.

But the kid started wailing.

"I looked at him, and I said, 'Why are you crying? There's nothing to be ashamed of. This is an injury that occurred in battle facing an enemy. You're going to get a Purple Heart. You're one of America's heroes.' "

But it wasn't that. The guardsman started begging Ling not to send him home. His work in Iraq was too important. " 'When I go home,' " Ling recalled him pleading, " 'do you know what I am? I work in a fast-food restaurant.' " "That just blew me away," said Ling. "That was my epiphany. If I could spend the rest of my life [helping] people like this—this nobody from California who's here doing what he believes to be this noble act—then, by golly, that's a life worth living."

When Ling returned from Iraq, it was clear he needed to ramp up the agency's neuroprosthetics effort. The agency's HAND program had already delivered some early successes, with researchers crafting brain-computer interfaces that enabled monkeys to control robot limbs and computer cursors. The research was promising, but it remained at the investigational stage. It wasn't the sort of thing they could deliver to young amputee veterans with the rest of their lives before them. For Ling, DARPA's neuroprostheses program needed to be more than a proof-of-concept showcase for brain-controlled prostheses. These vets had made huge, life-altering sacrifices for their country, and as he saw it, the country was duty-bound to return them to as normal a life as possible. "They represent guys like you and me. They're not there to rape and pillage. They are there to do something very noble," he said. "Dr. Tether, who'd visited Walter Reed, was really struck by these young service members with traumatic amputations. He saw a need for DARPA to get involved."

The result was that Ling would spearhead the government effort to create nothing short of a brain-controlled artificial human arm that had the shape, weight, and functionality of a biological limb—a "revolution" in upper-limb prosthetics.

•

It's not an obvious choice. After all, the vast majority of the estimated 1.7 million amputees living in the United States have lost all or part of a leg due to vascular complications associated with diabetes. Most amputees are older and less active than, say, a vet in her midtwenties who still wants to run, hike, or swim. For many amputees, their running and rock-climbing days are behind them. What they really want is a limb that fits comfortably and allows them to walk with a naturalistic gait.

And for the most part, that challenge has been met. "When you look into it, the lower extremity—that is, the leg prosthesis—was actually quite good," said Ling. "All you need is a hip that goes up and down, a knee that goes forward and back, an ankle that goes up and down, and a big toe that goes up and down—

just those four joints and you're very functional." This isn't to say an effective artificial leg is a simple engineering matter, but in the last fifteen years prosthetists have developed a range of specialized and highly functional lower-limb prostheses.

Just consider the disgraced South African sprinter Oscar Pistorius, who had both legs amputated below the knee after a congenital birth defect caused him to be born without fibulae. Before winning two gold medals and one silver in the 2012 Summer Paralympics in London, Pistorius had been banned from competing in the 2008 Summer Olympics for fear that his prostheses—a pair of curved carbon fiber "blades"—would give him an unfair advantage over his able-bodied competitors. Other leg prostheses abound. The German firm Ottobock introduced the C-Leg in 1997, an above-the-knee artificial leg that uses microprocessors at the joints to adjust the leg's gait, range of motion, torque, and ground orientation. A host of specialized lower-limb prostheses have since hit the market. Many of them, like the C-Leg, are essentially sophisticated computers built in the form of a leg, but there are also lower-tech devices, like flippers for swimming or spikelike pegs for rock climbing.

Artificial arms, by contrast, have remained stubbornly low-tech, progressing only incrementally since the third century when the Roman general Marcus Sergius had an iron fist crafted for himself after losing a hand in the Second Punic War. Static upper limbs remained the rule for the next seventeen hundred years, as the poor made do with crude hooks and the wealthy fashioned wood, leather, and iron into the shape of a hand.

Most arm prostheses were merely aesthetic, though some locked into shields, allowing knights to return to battle. The first known mechanical arms didn't arrive until the early sixteenth century, when a German mercenary named Götz von Berlichingen lost part of his right arm during a siege of Landshut. Made of iron and attached to the stump by leather straps, Berlichingen's hand had jointed fingers he could set to grip weapons. He used his iron fist for the next forty years, as he hired out his sword to several dukes, led rebels in the German Peasants' War, and later served Charles V during his campaign against the Ottoman Empire. Johann Wolfgang von Goethe later immortalized Berlichingen's life in an eponymous drama he based on the soldier's memoirs.

Other mechanical hands followed, but overall the technology remained stagnant. Surgery, a gory affair then dominated by the thinking of the second-century Greek physician Galen of Pergamon, did little to help the matter. Galen's often fantastical writings about human anatomy and disease were treated as gospel by physicians. But while Galen correctly surmised a difference between sensory and motor nerves, and he realized there was a distinction between arteries and veins, his work was also riddled with inaccuracies. He believed, for instance, there were two types of blood. He held to the Hippocratic notion of bodily humors, and he believed blood formed in the liver.

These errors echoed across the centuries, as anatomical and medical experimentation languished through the Middle Ages. Surgeons had very little understanding of germ theory or wound care, and their crude surgical methods ensured amputations carried an abominable 80 percent mortality rate. It was common practice for surgeons to remove damaged limbs with a guillotine or ax, later "detoxifying" the wound by cauterizing it with boiling oil or hot irons. Fever often ensued, and many amputees later died from massive hemorrhaging or septic shock.

These were the techniques that the sixteenth-century surgeon-barber Ambroise Paré learned while studying anatomy at the Hôtel-Dieu, the renowned Parisian teaching hospital. Paré might have continued practicing these methods had he not run out of the boiling oil solution he used to cauterize wounds during the Battle of Turin. Loath to leave soldiers' wounds untreated, the surgeon created a balm out of egg yolks, rose oil, and turpentine. He was surprised to discover the next morning that patients treated with his salve were recuperating well, while those who'd received hot oil were feverish and suffering "great pain and swelling about the edges of their wounds."

It was a revelation for Paré, who began questioning other medical practices as well. He soon realized that he could work to preserve the surrounding tissues during surgery, minimizing damage and aiding in recovery. He began experimenting with tourniquets during amputations, becoming the first surgeon to remove limbs with an eye toward crafting a stump to fit prostheses, and he later designed a prosthetic hand of his own. Known as Le Petit Lorrain, the mechanical hand boasted movable fingers and a fixed thumb.

But these mechanical arms, novelties really, were ahead of their time, and while surgical advances meant amputations became less deadly, upper-limb prostheses remained of limited use.

It wasn't until the early nineteenth century that the first truly movable prosthetic arm made an appearance. Developed by a German dentist named Peter Ballif, the artificial arm and hand used muscles in the back and shoulder girdle to extend and retract fingers. One century later, in 1912, a sawmill worker named David Dorrance created the body-powered split-hook-and-cable prosthesis after losing his right hand. Dorrance's design, which enables amputees to use their opposing shoulder to maneuver the arm and control the pincerlike hook, was the first to give amputees dynamic control of an artificial arm.

In the mid-twentieth century, scientists began to develop so-called myoelectric prostheses, motorized artificial arms that used electrodes to harness the tiny electrical impulses generated by the stump's remaining muscles. These arms gave users modest control, enabling wearers to open and close the hand by flexing their remaining arm muscles. Myoelectric limbs have since become more sophisticated. Nevertheless, the prostheses are often heavy and difficult to master. They generally have a limited range of available gestures, and many amputees prefer the hook-and-cable prosthetic Dorrance designed a century ago.

One reason artificial leg technology has flourished while arm prostheses have remained unchanged comes down to market size. Of the roughly 1.7 million amputees living in the United States, only an estimated 100,000 have lost an arm. Upper-limb amputations usually result from traumatic injury, and while veterans of the wars in Iraq and Afghanistan have swelled these numbers, they've done so only modestly: by the end of 2012, an estimated 1,700 service members had lost limbs during the wars. With such a small potential market, few companies have been willing to spend the sort of money it would take to overcome the formidable technical challenges presented by an upper-limb prosthetic.

Whereas the leg needs only in-line movement to walk, the human arm is called upon each day to perform a diverse lexicon of actions. Not only do the shoulder and wrist joints move side to side and up and down, but the upper and lower arms can also roll. Add to that the elbow, and a human arm needs seven degrees of freedom, or DOF, to function properly. The leg, by contrast, has six degrees of freedom (three in the hip, one in the knee, and two in the ankle). What's more, unlike an artificial leg, which during active use spends much of its time in a weight-bearing position, an artificial arm simply hangs from the stump. Lacking the supporting musculature of its biological counterpart, an artificial arm, no matter how light, dangles from its wearer like a dead weight. Lastly, the sheer breadth of extemporaneous actions the arm performs each day means that a truly "smart" prosthetic limb would need to have an elaborate yet seamless control mechanism.

This isn't to downplay the complexity of artificial legs, but there is a reason artificial leg technology has flourished over the years while most upper-limb amputees are stuck with early-twentieth-century technology. The arm is an order of magnitude more complex, and that's to say nothing of the hand, which is arguably the body's most complicated anatomical structure. "Your thumb already has as many joints as your entire leg, so that gives you a flavor of the complexity of the situation," said Ling. "It was clear that this was an area that required a tremendous amount of science—both basic science and basic engineering, as well as translational science to the patient."

It's easy to forget just how extraordinary hands are. When we wake each morning, many of us first reach to turn off an alarm clock. We rub our eyes, place our hand on the side of the bed, and begin the scooping, grabbing, tucking, pouring, and pressing of buttons that is morning ritual. We transition seamlessly from tasks like pouring cereal and slicing bananas to squeezing toothpaste onto a narrow row of bristles and combing our hair.

It's utterly mundane stuff. But from an engineering perspective, it's simply breathtaking: a multi-jointed anatomical structure that moves freely in space while executing a vast repertoire of unique movements.

The thumb may have as many joints as the leg, but the rest of the hand is no less intricate. The hand's twenty-seven bones form an elaborate network of joints, ligaments, muscles, and tendons. The muscles that control the hand reside mainly in the forearm, narrowing into tendons that pass through the wrist and connect to bone. Both sensory and motor nerves also tunnel through the wrist, branching out to coordinate muscular action and making the hand one of the body's most sensitive structures. The neuromuscular system between the brain and the hand is so deeply integrated that many actions—recoiling from heat, gripping an object more tightly as it slips from grasp—are often reflexive, occurring in the peripheral nervous system before conscious thought comes online. This delicate architecture is sheathed in a protective layer of subcutaneous fat and a thick, pliable skin that not only allows for less slippage while gripping but also pads the hand's inner workings.

But perhaps nothing speaks to the hand's complexity and central importance to our species so much as the amount of neural real estate it commands: the motor cortex devotes as much space to the hand as it does to the arm, leg, and trunk combined. With such a wealth of neurons, it's little wonder we can learn intricate tasks like playing the piano, sewing, or typing while also divining an object's identity just by holding it.

"It was thought to be an almost intractable problem because of the complexity of what it would take to get a fully functional arm from a robotic standpoint, but also, how do you control the thing?" Ling said. "In our opinion, the only way you could do that is to go directly into a process that is analogous to what we do naturally—that is, your brain. This became what I considered to be a DARPA-hard problem."

•

What followed was the formation of the Revolutionizing Prosthetics program, the DARPA-funded project that would tap the expertise of neuroscientists, electrical engineers, materials scientists, chemists, immunologists, physicists, biologists, and biomechanical engineers from more than thirty labs. The program's aim was ambitious. Not only would it deliver the world's first neurally controlled prosthetic arm that had the look, feel, performance, and weight of a natural human limb, but it would also be produced in a matter of years.

As Ling saw it, these kids had lost limbs fighting for their country. They were still young. They were still physically active. Some wanted to return to active military duty. Others wanted a normal civilian life. Either way, they had their whole lives in front of them. It wasn't so much that they needed the arm now. They needed it yesterday. "This may be a criticism of DARPA, but I think it is one of our strengths—and that's that we keep our eye on the goal," said Ling. "Everybody who joins the program from day one knows what the goal is, and they know what their part in that goal is."

In typical DARPA fashion, the Revolutionizing Prosthetics program tackled the problem on two fronts. The first, known as Revolutionizing Prosthetics 2007, was tasked with using existing technologies to create a quick and dirty limb inside two years. The program's second prong, Revolutionizing Prosthetics 2009, asked researchers to create a prosthetic limb from technologies they had yet to invent. "The arm had to be modular. It had to have all of these degrees of freedom. It had to look like an arm. It had to weigh like an arm," said Ling. "Then our dream was to have the brain directly control it as it would its native limb."

DARPA ended up awarding $18.1 million for the two-year program to DEKA Research and Development Corporation, a New Hampshire–based technology company headed by Dean Kamen, a legendary figure in engineering circles. Kamen was the driving force behind devices like the Segway Human Transporter, a two-wheeled scooter whose gyroscopes enable riders to control the machine with subtle foot movements, and the iBot, a self-balancing wheelchair that can rear up on its back wheels, allowing users to mount curbs, climb stairs, and "stand" at eye level with ambulatory people.

Kamen's team began with some of the myoelectric arms already in use. One of the great shortcomings of myoelectric prostheses, which rely on electrodes to pick up tiny electrical signals from muscles, is that there are often too few muscles in an amputee's stump to provide adequate information beyond a few basic gestures. The engineering challenge is one of supply and demand. Say, for instance, that an amputee has lost his arm above the elbow. He could use electrical signals from his biceps to close an artificial hand and signals from his triceps to flex the elbow. But if the only two muscles remaining in the stump are already spoken for, how can engineers tease out additional information to control the twisting of the lower arm or the movement of individual fingers?

Some researchers have sought to work around this by programming different switches into myoelectric limbs and training users to control the arm's complex muscle contractions (flex your biceps twice to bend at the elbow, that sort thing). But with such a cumbersome and limited muscular vocabulary, these workarounds often provide users with only a few more gestures, and many amputees mothball their myoelectric arms in favor of the old split hook and cable.

Recently, however, a researcher named Todd Kuiken had pioneered a novel surgical technique known as targeted muscle reinnervation, or TMR. During surgery, Kuiken, who directs the Center for Bionic Medicine at the Rehabilitation Institute of Chicago, would splay the arm's nerve bundle at the amputee's stump, transferring its individual strands to muscles across the chest. The frayed nerves were still devoted to specific arm gestures, like flexing the elbow and moving the wrist. As the nerve fibers eventually grew into the chest muscles, users could contract specific areas of the pectoral muscle when they thought about those particular movements. With the nerves spread across the chest, TMR not only gave prosthetists a broader range of control signals (and a greater repertoire of robotic commands) but also endowed users with a more naturalistic interface, enabling them to link nerves that once controlled, say, grasping with the hand to a similar action in the prosthetic limb.

Incorporating Kuiken's work, Kamen's team produced a working prototype within their two-year time frame, completing the so-called DEKA arm in 2007. The work persuaded DARPA to continue funding the project, and DEKA engineers have since fine-tuned the arm, running clinical trials and producing new generations of the arm before submitting it for FDA approval—the final step before offering it commercially. The FDA approved the DEKA arm in 2014, clearing the way to make it the most sophisticated artificial upper limb on the market.

Nevertheless, the DEKA arm prototype, made on the fly from existing technologies, had only ten degrees of freedom (compared with the twenty-seven DOF in a human arm and hand). The hand prototype had six preprogrammed grips, and the arm was not a direct neural interface; it could be controlled either by myoelectric sensors or by foot pedals. "It's not the neural-controlled arm, I accept that," said Ling. "Still, think about something that went from conception to FDA submission in five years. Name me another medical product at this scale that has done that, and they made it de novo too. It's not like we got shortcuts. And we did it in *five* years! And I'll tell you why we did that. It's because our eye was constantly on the mark."

But the DEKA arm was only the beginning. The project's complementary second prong, Revolutionizing Prosthetics 2009, aimed to completely reimagine upper-limb prostheses, creating an anthropomorphic arm that had the weight, size, and functionality of its biological counterpart. The prosthetic would need the strength of a natural human arm. It required a similar range of motion. Users should be able to gain fluid control over the limb, moving it spontaneously as opposed to performing a set of preprogrammed gestures. Ultimately, the prosthetic would need to convey sensory feedback to the wearer, registering not only pressure and heat but also a sense of proprioception, allowing users to know the arm's position in space. "It was a bold thing to ask," said Ling. "The other part was, how do you properly control it? But our dream, which is to actually have their brain directly control it as it would their native limb—that required a tremendous investment in the brain-machine interface, a huge investment into basic neuroscience."

DARPA awarded the initial $30.4 million contract to the Applied Physics Laboratory, or APL, a longtime national defense contractor affiliated with the Johns Hopkins University. The lab was first organized during World War II as part of the government's push to harness the scientific and engineering expertise of the nation's universities to further the war effort. Though APL's work was less visible than the Manhattan Project, its scientists developed a critical device that more accurately detonated antiaircraft missiles.

At the time, the only detonation devices were either timed fuses, which detonated a warhead at a specific interval after firing, or contact fuses, which detonated upon impact. These earlier fuses were all but useless against the new breed of agile fighter planes that ruled the skies. Working out of a converted auto dealership, APL scientists developed the variable-time proximity, or VT, fuse, a device that used a radio transmitter and receiver to detonate warheads as they approached a target. The fuses were tremendously effective at protecting U.S. warships from dive-bombers in the Pacific and Allied troops against buzz bombs at the Battle of the Bulge. The VT fuses gave Allied forces such a technological advantage that APL engineers outfitted them with a self-destruct mechanism to ensure dud fuses didn't fall into enemy hands. "When all armies get this shell we will have to devise some new method of warfare," General George S. Patton quipped. Along with the atomic bomb and radar, the VT fuse was one of the crucial weapons advances during the war leading to the Allied victory.

APL went on to win innumerable classified and nonclassified government contracts after the war. The lab has developed everything from guided missile systems for the navy and satellites for NASA to automated tracking radar systems and missile defense technologies. The lab's work has also underpinned civilian technologies like the moving walkway and the rechargeable pacemaker.

With its deep bench of engineers, long history of government contracts, and sizable resources, APL was a natural choice for DARPA. Even so, the Hopkins group enlisted some thirty labs outside its Silver Spring campus to complete the arm. At its height, the team comprised more than four hundred individual researchers, including electrical and mechanical engineers, software and signal analysis specialists, wireless communications experts, and authorities on human form factors, cosmesis materials, reliability, and manufacturing.

The science was no less complicated, and APL recruited a team of neuroscientists that specialized in neural motor decoding, neural stimulation, and sensory feedback. And that's to say nothing of the surgeons, the prosthetists, and the physical and occupational therapists who would eventually work with research subjects. "The brilliance of what APL did was to see that to solve this problem, it wasn't in a better algorithm to interpret what the neural patterns mean, and it wasn't in a better electrode to put in the brain, or a better prosthetic limb—it was all of those things," said Mike McLoughlin, an electrical engineer at APL who took over as the project's administrator in 2009. "You have to bring all of those disciplines together, but you really have to orchestrate it in a way that it's coming together at this single point and not just everybody going off and doing their research."

The arm had to be unlike any that had come before it. Not only would it need to have the look, weight, and size of a human arm, but it would also need to move naturally. It had to curl twenty kilograms at the elbow, squeeze thirty-two kilograms in its grip, and sport a protective sheath that didn't cause its motors to overheat. It would also require an intuitive control mechanism—one that didn't require users to learn a glossary of preprogrammed commands.

The limb, in a word, would have to be integrated into the wearer's nervous system. But just as they needed to wed the prosthetic to the body's motor neurons, they also needed to craft a limb that could deliver sensory feedback. What's more, the arm also had to be reliable and comfortable—a tall order for what amounts to dead weight dangling from the amputation site.

The design challenges didn't stop there. The arm would require a rechargeable, onboard power supply. The prosthetic would also have to be modular, meaning not only that it could be modified to accommodate amputees who had lost an arm at the shoulder, elbow, or wrist but that each module would house its own digital smarts. That was a challenge in the upper arm and forearm. But it was the hand, with its nineteen degrees of freedom, that seemed next to impossible.

The more elegant engineering solution would be to construct the prosthetic hand along similar principles as its biological counterpart, with its power in the forearm. But such a limb wouldn't serve an amputee who'd lost his hand at the wrist. To be truly modular, the APL hand would need to house all of its microprocessors and motors in the palm and fingers. This meant not only that engineers would have to drastically miniaturize their technologies but also that the hand might be disproportionately heavy in relation to the rest of the arm.

It was a struggle between the hand's size, its weight, and its power. "Getting any two of those is easy," said McLoughlin. "It's when you throw in the third that things get really tough." Motors had to be shrunk, electronics miniaturized, and circuit boards compressed. APL's engineers were cramming very complicated, de novo technologies into very small spaces, pushing the state of the art of microprocessor circuitry and mechanics. "That was a real challenge," said McLoughlin. People aren't used to fabricating on that scale."

Working small created other challenges as well. Engineers were so focused on maintaining the arm's power while keeping its weight down that the arm's network of tiny motors could easily overheat. The arm lacked the mass to dissipate heat effectively, and engineers had to keep a close eye on the arm's power usage to ensure it didn't malfunction.

Some labs, like the one at Oak Ridge National Laboratory in Tennessee, inevitably fell away. The Tennessee lab had been working on a micro-fluidics technology to move the arm. The miniaturized hydraulic system would vastly reduce the arm's weight and power consumption, but the technology wasn't ready. "They did some really nice work," said McLoughlin. "But with the time frame we were on, it just wasn't something that was mature enough to be able to integrate."

This great sifting operation consumed the APL team for almost two years, as scientists weighed competing technologies. Meanwhile, they were also looking at different control methods. Could they master the arm with a simple EEG electrode cap? Could they tap into the peripheral nervous system, using the residual nerves at the stump? Or should they go directly to the source, giving the brain direct control over the limb?

Like Kamen, the APL project wrapped up this first phase by recruiting Kuiken's group out of Chicago. Linking an early prototype to a research subject's reinnervated pectoral muscle, the APL team persuaded DARPA to continue funding the project, which enabled them to build another prototype.

DARPA may be very good at developing technologies for future applications, but it's not in the business of transitioning those technologies to the marketplace. It relies on industry for that. The hope was that once the APL team had produced a neurally controlled anthropomorphic arm, the private sector would step forward to take the arm through FDA approval and commercialization.

But again, they ran into market size. "There are not a lot of amputees in the world," said McLoughlin. "Nobody is going to come in and put tens of millions of dollars into the research and development of the arm when there is a very small market."

Private investors balked, but DARPA decided to fund an unplanned third phase of the program—one that would work out some of the kinks and finally demonstrate direct brain control in humans. At the time, only a handful of neuroscientists had shown themselves capable of leashing the mind to a prosthetic limb. And those who had were working almost exclusively in rats and nonhuman primates.

It was an entirely new challenge, and the list of scientists with the neuro-chops to pull it off was short. "We looked at all the proposals," said Ling. "Andrew Schwartz understood what we were trying to do. His proposal was singular in its determination to meet this goal—that is, getting a human to run this arm in a multifunctional way, dexterous movement, as Andy likes to say. He is singularly minded, that guy."

3. MONKEY MAN

Though he died in 2007, Matthew Nagle looms over the field of neuroprosthetics, embodying both its limitations and its promise. Once a standout high school athlete, Nagle was paralyzed from the shoulders down while trying to help his friends in a fight. It was July 3, 2001, and Nagle had gone to Wessagusset Beach, an inlet south of Boston, to watch the fireworks. Burly at six feet two inches, Nagle still held the record for unassisted tackles at Weymouth High School. He was also a scrapper, and when his friends found themselves embroiled in a fight, Nagle plunged into the scrum, working his way through the churning bodies toward his friends. As Nagle fought to help his buddies, he encountered Nicholas Cirignano, a twenty-year-old who wielded an eight-inch hunting knife.

What happened next remains unclear. Nagle said he had no memory of the attack, but in the roiling melee Cirignano sank his blade into Nagle's neck, severing the spinal cord and rendering him paralyzed.

He would never walk again.

Nagle was unbowed, however, and in June 2004 he took the audacious step of becoming the first human research subject for the BrainGate Neural Interface System, a BCI being developed by Cyberkinetics, a neurotechnology company founded by the Brown University neuroscientist John Donoghue.

Nagle was not the first human to be implanted with electrodes. That distinction goes to John Ray, a Vietnam veteran who was left completely paralyzed after a brain stem stroke. Led by Philip Kennedy, researchers at Emory University had implanted Ray with a pair of glass-encased electrodes, known as neurotrophic electrodes. Using only a few neurons, Ray gained rudimentary control over a computer cursor and was able to type brief messages.

But Nagle was different. Whereas Ray had just two electrodes, Nagle received a Utah array, a pill-sized implant whose ninety-six microelectrodes bristle from its base like a bed of nails. Donoghue wasn't the only neuroscientist recording from scores of neurons at the time. Andrew Schwartz, Donoghue, and a third neuroprosthetist, Miguel Nicolelis, had been in fierce competition for years, recording multiple electrodes in monkeys. The difference was that Donoghue was recording in a human.

•

By almost any measure, Andrew Schwartz had by then already met with great success. He'd published several well-received papers on how he had endowed monkeys with direct cortical control of machines. The University of Pittsburgh had recently lured him away from the Neurosciences Institute in San Diego with promises of a handsome brain lab and the capacity to graduate from the monkey cortex to the elegant realm of the human neocortex.

Nevertheless, by 2004 Schwartz felt he was falling behind. His rival Nicolelis had floored colleagues a few years earlier when he landed a $26 million DARPA grant. Not only had Nicolelis embarked on a series of headline-grabbing experiments—enabling monkeys to gain neural control over robots and robot arms—but the flamboyant Brazilian's enthusiasm for the field was irrepressible. He was given to grand pronouncements about the coming cyborg age, making him a favorite of the science press. Other researchers might have grumbled that Nicolelis's work could be messy or glib, but the fact remained: he was out front, and his visions of a neurally connected future were catnip to the popular imagination. Now flush with DARPA cash, Nicolelis seemed poised to dominate the field.

If that weren't bad enough, here came the avuncular Donoghue—a tweedy East Coast Apollo to Nicolelis's Latin Dionysus. Like Schwartz and Nicolelis, Donoghue had been working in monkey cortices for years. But whereas Nicolelis was a showman—deliberately pushing the boundaries of how we think about the brain, its neural plasticity, and our complicated relationship with technology—Donoghue presented himself as a sober researcher-cum-businessman. Cyberkinetics was his great play. He intended to create a marketable BCI for the disabled, and by pressing his Utah array into the brain of Matthew Nagle, Donoghue not only presented himself as the field's courtly emissary but also set himself apart from his rivals, becoming the first researcher to implant scores of electrodes in a human.

"I was scared shitless," said Schwartz. "I thought Donoghue was going to knock my socks off." Schwartz had been a careful scientist his entire career, performing well-planned studies that were precisely executed. His data were rock solid. He made few tall claims, and he was uneasy around a popular press that loved the sizzle of neuroprosthetics—the promise of putting Google in our brains—but had little interest in the basic science.

But basic science was what made a guy like Schwartz tick. He had devoted his entire career to understanding the neural underpinnings of arm movement, and he thought sweeping claims about the field's future were a sort of scientific heresy. BCIs were a means toward understanding the brain. Perhaps neuroprosthetics could eventually help people with spinal cord injuries, but the notion that BCIs were anything but a research or rehabilitative tool wasn't merely premature for Schwartz. It went against everything he believed. "That to me is just wrong. It's almost immoral. It's, like, nonscience," he said. "I just refuse to go there. I always say,

'You know, that's science fiction, and there are other people a lot better at science fiction than I am. Go talk to them.' These guys are nuts."

Schwartz continued to attract money for his monkey research, but he knew he'd have to work with humans to stay competitive. Now Donoghue had the jump on him. Cyberkinetics was amassing millions in venture capital, and Nagle, who was hailed by *Wired* magazine as the first "neuro-cybernaut," was just the beginning: the FDA had granted Donoghue license to implant many more human patients.

"If he could do the same kind of stuff with his human patients that I was doing with my monkeys, I would have become irrelevant—instantly," Schwartz said.

Nagle kept his implant for thirteen months, during which time he used it to control a host of computer-based devices, playing the video game *Pong* and using a cursor to open e-mail, play MP3s, and draw geometric forms.

Meanwhile, Donoghue's team recruited a second patient, and by 2006 they published their results in the journal *Nature*. The journal's editors gave Donoghue and his colleagues the Cadillac treatment, publishing Nagle's photograph on the cover and penning an accompanying editorial along with two news stories. The popular press soon followed. "If your brain can do it, we can tap into it," Donoghue triumphantly told *The New York Times*, following up by boasting to CNN that his research represented the "dawn of the age of neurotechnology."

Donoghue's colleagues were more reserved. By then, both Schwartz and Nicolelis had demonstrated that monkeys could gain neural control of computers. By that measure, his colleagues said, Nagle's unsteady command of a cursor was not terribly impressive. Using an average of twenty-six neurons, it took him 2.5 seconds to move the cursor from the center of the screen to an icon at the edge. He completed the task between 73 and 95 percent of the time. By contrast, it would take a biologically intact person using a mouse a fraction of a second to move that distance, and she would have something approaching 100 percent accuracy. "If you are going to have something implanted into your brain," Jonathan R. Wolpaw, an EEG researcher at the New York State Department of Health, told the *Times*, "you'd probably want it to be a lot better."

The study's second patient, an anonymous fifty-five-year-old man with a C4 spinal cord injury, was less successful. His implant had problems with the electrical contacts in the pedestal that exited his skull. After making a repair, Donoghue's team began recording some seven months after the implantation surgery, managing to record from around fifty neurons. Three months later, however, the patient withdrew from the study when most of his electrodes stopped picking up neural activity.

Nevertheless, the study was a watershed moment for the field, striking a deep chord with the public and positioning Donoghue as the face of neuroprosthetic research. Meanwhile, they'd also managed to win FDA approval, and they had demonstrated unequivocally that a sophisticated brain-machine interface could work in a human—major accomplishments that no one in the field could deny. "They put it in a human, and they showed the guy lived, and they got units," said Schwartz. "I give that to them. I give it to them today. They got it."

•

At some essential level, the brain's sole function can be distilled to one task: issuing motor commands. Whether the brain is directing the lungs to breathe, the heart to beat, the hand to write, or the mouth to speak, the only means we have to express our thoughts and affect the outside world is through muscular activity—be it a blink of the eye, a wave of the hand, or a balling of the fist. Thoughts divorced from movement remain trapped in a sort of mental purgatory. They are messages composed but never sent. They cannot be shared with the outside world, and in some critical manner they fail to exist. "To move things is all that mankind can do," the English neurophysiologist and Nobel laureate Charles Sherrington once told a crowd at Cambridge University. "For such the sole executant is muscle, whether in whispering a syllable or felling a forest."

More recently, neuroscientists have taken this idea even further, arguing that the only reason we evolved brains in the first place was to produce physical movements—actions we could adapt to suit life's shifting conditions. On an existential level, organisms need thought (or at least some thought-like process) for their survival: we need thought to seek out food sources and avoid predators. And while our endowment to recall life's more subtle experiences may be a defining aspect of human intelligence, there can be little doubt that those delicate reflections are but the rarefied descendants of brain functions that enable us to remember hidden food sources and how best to elude predators.

As proof of the concept, neuroscientists such as Daniel Wolpert point to the sea squirt, an immobile filter feeder that spends its days attached to a reef while mindlessly trawling for nutrients. What makes the animal remarkable is its metamorphosis to reach this state. The sea squirt does not begin life attached to a reef. Rather, it swims freely, seeking out food by sifting nutrients it encounters in the open water. Once the animal matures, however, the sea squirt attaches to a reef where it will remain the rest of its life. And here things get interesting: the animal's first order of business is to digest its brain. The organism no longer needs to move about in search of food. It no longer needs to coordinate muscular activity. And without the need to move, the animal no longer needs its calorie-hungry brain.

In other words, the brain, its thoughts, memories, and intentions—our conscious experience of the world—is intimately tied to physical actions and muscular activity. At least that's the theory, and recently researchers at Carnegie Mellon University in Pittsburgh have embarked on a series of experiments that seem to bear it out.

One of the great recent advances for peering into the brain is called functional magnetic resonance imaging, or fMRI. Developed in the early 1990s, fMRI charts brain activity by measuring cerebral blood flow, relying on the assumption that increased blood flow indicates heightened neural activity. The technology, which allows researchers to observe continuous brain activity in successive snapshots, enables researchers to observe how the brain responds to everything from thoughts of love and metaphors to classical music and meditation.

Words are no different, and scientists have used fMRI to detect how specific areas of the brain "light up" when a person thinks of a particular word. In one study, the researchers Tom Mitchell and Marcel Just asked test subjects to concentrate on some sixty nouns while undergoing fMRI scans. The researchers then analyzed the nouns to gauge how often they occurred in conjunction with twenty-five verbs associated with sensory or motor function—verbs like "see," "eat," "push," or "drive." By comparing their statistical analysis with the individual fMRI data, Mitchell and Just found they could predict that specific patterns of neural activity would emerge when people thought of particular objects. In essence, the researchers could determine (with a mean accuracy of 77 percent) when people were thinking of things like airplanes, buses, or apples.

This sort of "mind reading" is intriguing in and of itself, but perhaps more important was the method Mitchell and Just used to forecast which objects people were thinking of. They used verbs. The scientists found they could divine, say, that a person was thinking of an apple by looking at the sensory and motor areas of the brain associated with how a person might hold, bite, or taste an apple. The implication is tremendous: we don't merely perceive physical objects with our senses; we actually represent those objects as things to be physically acted upon, held, touched, or eaten. "We are fundamentally perceivers and actors," said Just, commenting on the research. "The brain represents the meaning of a concrete noun in areas of the brain associated with how people sense it or manipulate it. The meaning of an apple, for instance, is represented in brain areas responsible for tasting, for smelling, for chewing. An apple is what you do with it."

Physical activity and moving muscles are not concerned only with how we interact with objects: they are fundamental building blocks we use to make sense of the world.

•

"Look at those units!" Schwartz barked as he made his way toward a workstation at the far end of his lab. Walking past several freestanding computer consoles that rose like stalagmites from the floor, the scientist was moving toward something indiscernible to the uninitiated. To a man like Schwartz, however, it was nothing short of awesome. "They're huge!" he cried.

Along with the oscilloscopes, video monitors, routers, speakers, and desktop computers, the workstation housed several racks of neural processing units. The black boxes were each about the size of a stereo amplifier. They were connected to thick wires that hung like vines from the lab's ceiling, linking the workstations to the heart of the lab—a suite of monkey bays where over the years Schwartz has demonstrated some of the best neural control in the business.

Months earlier, Schwartz and his colleagues had implanted several Utah arrays in the motor and sensory cortices of a monkey named L1. That monkey was now in one of the bays playing a video game. As the animal worked, electrodes in its sensory cortex delivered tiny pulses of electricity, enabling L1 to perceive tactile sensations from the game. A speaker atop the workstation crackled with static noise. It sounded like rapidly popping popcorn or snow on a television. What it was, in fact, was the chattering hum of L1's neurons as they volleyed information back and forth—the neural code, raw, physical thought that's incomprehensible to the human ear.

The computer, on the other hand, was having an easier go of it. Each crackling neuron appeared as a wave within a graph. Each wave followed the same pattern. When a neuron spiked, a pulse of electricity would run the length of the cell. On the screen, this looked like a typical neural signal: a dramatic spike at the front, followed by a sharp drop, which promptly leveled out. It was about as close to observing an intact neuron as you could get.

Graphs like this are ubiquitous in brain labs. But these signals? "They're huge!" Schwartz cried again.

More surprising, though, was the signals' longevity. Schwartz and his colleagues had implanted L1 several months earlier. The signal quality should have begun to degrade by now, as the body's immune system cordoned off the offending intruders, placing more distance between the sensor and the cell and dampening the electrode's sensitivity. But Schwartz and his colleagues had treated L1's arrays with an experimental immunosuppressant. Months later, the signal had weakened, but only slightly. "The electrodes are close," he said. "Right now we get two to three years' use out of them. It's not going to take that much tweaking to get them up to five to ten years."

Schwartz is a slight man and balding. He has a broad, leonine nose. His hazel eyes can seem impenetrable behind his thick glasses, and his graying hair gets a little shaggy on the side. He walks with a jogger's gait, stepping lightly on the balls of his feet, and he dresses simply: parkas and sweaters in winter, sandals and tube socks in summer.

His lab looks like a tinker's workshop, as a parade of neurally implanted monkeys circulates through the room on rolling Plexiglas task chairs. Against a far wall stands a worktable littered with spools of multicolored wires. Microscopes stand nearby, as do several bottles of rubbing alcohol. Task lamps cling to the edge of a table, where drills and vises sit below shelves housing everything from WD-40 and Crisco to bins titled "Burrs & Bits" and "Force Sensor Resistors." A broken toy scimitar is crammed between the bins of one shelf, while another shelf, titled "Team Robot," holds a hodgepodge of power connectors, oscilloscopes, desk clamps, and soldering equipment.

In a closet by the entrance, Schwartz keeps what amounts to a minor museum of robot arms. On one shelf, opposite monkey treats like graham crackers and Lucky Charms, hulks a superannuated limb made of blue tubular metal. It has three joints and, from the looks of it, hasn't been used in years. Thick green cables spill like spaghetti from its base, and a fork sprouts hook-like from the wrist.

Meanwhile, the so-called Shanghai arm sits in a box on the floor. When he bought the arm in the late 1990s, multi-jointed robotic limbs were hard to find in the United States. Schwartz looked everywhere, eventually finding a company in Shanghai. But these were the early days of the Internet, and it wasn't clear how he could buy it. "They wouldn't ship the arm unless they had the money," Schwartz recalled. "I didn't want to give them the money unless I had the arm. So I ended up going there with a check for something like $70,000."

Schwartz brought a suitcase to China, planning to return with his prize. But the deal fell through. He needed an export license, so Schwartz flew home empty-handed. "At least I had seen the arm. I knew it existed," he said.

As it turned out, the Shanghai arm was probably more trouble than it was worth. The gears were all made by hand. Whenever one broke, which was often, Schwartz had to call China for a new one. After a while, the Chinese manufacturers didn't want to make the gears anymore, so Schwartz began collecting old Shanghai arms for parts. Still, the pieces weren't interchangeable. "Each piece was changed for each model," he said. "It was really a pain in the butt."

•

Like most neural prosthetists, Schwartz has an origin story for the moment he realized the brain might someday control machines. He was working as a postdoc in Apostolos Georgopoulos's lab at the Johns Hopkins School of Medicine. It was the early 1980s, and Schwartz had just graduated with a PhD from the University of Minnesota, where he had specialized in the brain's role in motor output (giving his dissertation the somewhat forbidding title "Activity in the Deep Cerebellar Nuclei During Normal and Perturbed Locomotion").

Schwartz had been captivated by the riddle of movement since childhood, when he learned the spinal cord couldn't repair itself. The liver can regenerate. The skin heals with scar tissue, and individual cells can repair themselves. But the spinal cord? Once it's severed, it never heals. "I've been interested in that question ever since," he said. "Why the hell can't the spinal cord fix itself?"

When he was an undergraduate at the University of Minnesota, Schwartz's fascination with paralysis led him to a brain lab in the basement of the school's library. He had one question for the researcher who answered the door: Do you know anything about paralysis?

"He just laughed at me, said come back tomorrow," said Schwartz. "They took me on as a lark. You know: We'll give you a few rats. Now go in the corner and have fun, kid."

Over the next few months, Schwartz hatched an experiment he thought might lead to a cure for paralysis. He'd come across an obscure journal article that theorized an autoimmune response to spinal cord injury. Unlike the body's other organs, the central nervous system is sheathed with a dense layer of endothelial cells. This sheet of tissue, known as the blood-brain barrier, selectively prevents blood-borne substances from entering the brain. It ensures the neural environment remains constant by segregating it from neurotoxins that could destabilize its chemical balance.

The article conjectured that when the spinal cord is injured, the tearing of neurons and capillaries breaches the blood-brain barrier. Neurons try to repair themselves, but as they seek to bridge the gap and reforge connections, the body's immune system attacks the freshly exposed neural tissue as it would a foreign substance.

The idea was based on a feature of multiple sclerosis, where the immune system attacks a protein in the myelin sheath, a cellular membrane unique to the nervous system that increases neural communication speeds. "I really liked the idea," said Schwartz. "I was pretty naive." During his experiment, Schwartz exposed a small population of neonatal rats to the protein before their immune systems had fully developed. His hope was that their immune systems would learn to recognize the protein as part of the body and not as a foreign object. He let the rats mature; then he severed their spinal cords to see if they would recover.

They never did.

Nevertheless, Schwartz was hooked. At the time, the field of motor physiology was deeply influenced by Edward Evarts, the hardworking head of the neurophysiology lab at the National Institute of Mental Health in Bethesda, Maryland. Evarts's career was nearing its end. As a younger man, however, the Harvard-educated scientist had pioneered a groundbreaking technique for recording individual neurons.

Researchers had been using electrodes to measure brain activity for years, but accurate recordings remained hard to come by. Not only did scientists have to drill tiny holes in the skull, but they also had to drive the electrode into the neural mass. The trick was to descend the wire close enough to get robust recordings, but not so close that they damaged the cell. The brain's flan-like consistency didn't help matters. Simple movements like walking, sitting, and even breathing can jostle the brain, causing individual neurons to shift slightly in the skull case.

This natural movement may be minute, but so is the strength of brain signals. If an electrode is fixed or doesn't move naturally with the brain, it can lose contact with the recorded cell, causing the signal to change or even disappear. From a practical standpoint, these technological restraints meant that many invasive brain experiments were done on anesthetized animals. Researchers, careful not to disturb the electrodes, would manipulate the limbs of their unconscious research subjects while recording their neural activity—an inherently limited paradigm in that the animal is not conscious and initiating the movement.

Evarts's great technical innovation was to improve on a hydraulic drive system (originally designed by David Hubel) that used a piston and cylinder to sink microelectrodes to specific depths in the brain. The apparatus was attached to the skull and immobile, but the electrode could easily be extended and retracted. Evarts's micro-drive enabled him to record individual neurons in lab animals that were awake and moving, giving him a leg up on scientists who were confined to working with anesthetized research subjects.

The results were groundbreaking. At thirty-six years old, Evarts had upended previous neurological theory by demonstrating that some neurons are as active during the rapid eye movement phase of sleep as they are when an animal is awake and visually observing the world. It was an important finding, showing the brain was far from passive during sleep.

But it was Evarts's work in the motor cortex that dominated Schwartz's world. In the mid-1960s, Evarts began training his monkeys to perform specific physical actions when prompted by a visual or auditory cue. Evarts recorded neurons in the animals' motor cortex from the prompt all the way through the task's completion. By studying individual neurons in awake animals, he found that cells in the motor cortex changed their firing patterns about sixty milliseconds prior to muscle engagement. Perhaps most significant, Evarts found that the faster some neurons fired, the greater the force generated by the muscles in movement.

His findings pointed to a proportional relationship between motor cortical activity and muscular action. Evarts and his collaborators later found that the motor cortex appeared to use smaller neurons to initiate subtler movements involving less muscular energy but recruited larger neurons for movements requiring greater force. In essence, Evarts's research indicated the motor cortex used a two-pronged approach to control movement: it increased neural firing speeds and recruited larger neurons to create movements with greater force while using smaller neurons to initiate finer actions.

"This clicked. It was simple to understand because everybody thinks the motor cortex is hooked directly to the muscles," said Schwartz. "So it made sense that a neuron would fire more, just like you'd expect a muscle to do."

But this paradigm, like so many in neuroscience, was an imperfect conclusion based on incomplete data. It didn't take into account some of the more abstract features of movement; namely, Schwartz believed, it didn't account for context. For instance, if you bend your arm with the palm up, you have to engage your biceps to flex at the elbow. By contrast, if you extend your arm out to the side with the pinkie edge of your hand facing up, you have to engage an entirely different muscle group to bend at the elbow. The outcome may be the same, but the muscles contracted are entirely different. The question for Schwartz, then, was if the motor cortex was directly linked to muscles, how could it use the same neurons to coordinate action in different muscles?

"There is a relationship between motor cortex and muscles, but as you start doing something that's a couple steps removed from muscle contraction, it gets really complicated," said Schwartz, who as a postdoc with the Georgopoulos lab would begin to dismantle the accepted model. "Those motor cortical cells are still going to fire, but then there are all these different pathways [the signals] can go through to get to the arm. It's going through a maze, and there are many different ways it can go, depending on all these other factors."

Working with rhesus monkeys, the Georgopoulos group set up a classic "center-out" task using a central button encircled by eight "target" buttons. The experiment began when the center button lit up. If the monkey pressed the center button, one of the surrounding targets would light up. After the monkey successfully pressed the second button, it would receive a small sip of juice for its trouble. Meanwhile, Schwartz and his colleagues had implanted microelectrodes in the animals' motor cortices, enabling researchers to study nearly three hundred individual neurons while also using a visual tracking system to monitor the animals' physical arm movements.

What they found would upend the current thinking about neurons and muscle contraction. "When you're moving a 2-D lever, you might think simplistically that it's just muscles involved—that it's just an expression of muscle contraction," said Schwartz. "But when you go to 3-D space, that relationship becomes much more complicated. It starts looking like it's not just muscles—maybe it's something different."

Schwartz and his colleagues found that individual neurons weren't simply linked to specific muscular actions and directions. For instance, a neuron wasn't uniquely coded to the biceps, increasing its firing rate when it wanted to bend at the elbow but remaining dormant the rest of the time. Nor were neurons strictly linked to specific directions, firing intensely when the arm moves up and to the right but remaining inactive when the arm simply moves to the right.

Rather, the Georgopoulos group found that individual neurons were broadly "tuned" to a specific direction: they would fire at different frequencies depending on the particulars of a desired movement. For instance, an individual neuron might fire intensely to move the arm up and to the right, but that same neuron would fire, albeit at a different rate, when simply commanding the arm to move up. It wasn't that individual neurons were solely responsible for certain movements or linked only to a specific set of movements. It wasn't even that individual neurons contributed more to some movements than to others. What Schwartz and his colleagues found, rather, was that *populations* of neurons best determined unique movements. Each cell in a neural ensemble was "directionally tuned," meaning it would fire when initiating a movement in a certain direction.

Directionally tuned cells fire whenever the intended movement is in a broad general direction—firing more or less frequently given the particulars of the movement. Some cells might fire 250 times per second. Others might fire 100 times per second, while still others might fire only twice. It wasn't that neurons were active during specific activities and rested during others. Rather, neurons across the population were constantly modulating their firing patterns, speeding up for some gestures, slowing down for others.

They formed fleeting patterns that quickly dissolved, seamlessly evolving into new temporary constellations. Each cell fired in correlation to each action, and researchers found that by studying firing rates across a neural population, they could accurately re-create the trajectory of an animal's three-dimensional movement.

Previous studies had shown that by recording from populations of neurons, you could predict force. Similarly, Schwartz and his collaborators had used more neurons to predict two-dimensional movements, but this was different. "The question was, well, it worked once under these limited laboratory conditions. But now we have an animal reaching with his hand in free space, and the same principle held up," said Schwartz. "It starts sounding more like a law or a principle."

The controversial series of experiments coming out of the Georgopoulos lab took direct aim at the prevailing ideas about the motor cortex. Nevertheless, its basic scientific findings were so compelling that in 1986 one of the lab's articles made the cover of *Science* magazine, a sort of academic Mount Everest. The popular press soon followed, and it was then, during an interview with *The Baltimore Sun*, that Schwartz first seriously considered brain-machine interfaces.

"This was the first time that it was really conclusive that you could look at something as abstract as movement direction and get such an accurate readout," he said. "People were interviewing us about that, and they said, 'Well, what's this going to do for people?' We had to kind of pull our chins and say, 'Well, maybe this will help paralyzed people someday.' "

•

DARPA didn't select Schwartz's application for the Revolutionizing Prosthetics program in 2005. The agency instead selected the Hopkins team, which oversaw a host of labs working on the project.

"It was a complete nightmare," Schwartz said.

By then, he had firmly established his reputation as a monkey man. Three years earlier, in 2002, Schwartz had completed a series of important experiments where he granted a pair of monkeys neural control over a computer.

Still, it wasn't the first time a researcher had managed such a feat. Two years earlier, in November 2000, Miguel Nicolelis and his colleagues had stunned the scientific community when they endowed an owl monkey named Belle with simultaneous neural control over two robot arms—one in another room at Duke, and another six hundred miles north in a laboratory at the Massachusetts Institute of Technology.

Belle had used a joystick to move a cursor along a horizontal string of lights. As she moved her hand, the micro-wires implanted in her motor cortex ferried her neural activity through a battery of cables and algorithms (as well as the Internet) to command the two robot arms, syncing the limbs with physical movements she was making in real time. "I could only think of what Galileo Galilei had allegedly murmured in his own defense during his trial before the Italian Inquisition," Nicolelis later wrote of the moment the MIT arm went live. "*Eppur si muove*"—"And yet it moves."

The experiment was pure Nicolelis—both splashy and controversial. The Brazilian had managed to show he could decode neural activity in real time. He had shown that using the Internet, the brain could project its intentions not merely onto machines in the next room but to machines that were hundreds of miles away. With BCIs the brain was networked, the body an arbitrary boundary. On the other hand, Nicolelis had used an owl monkey, a primate so small it can fit in your shirt pocket and thought by many to be a less reliable animal model than the larger rhesus monkey.

Nevertheless, Nicolelis got there first, or at least that's how it was portrayed, a bit of showmanship that still burns Schwartz today. "We had a paper in 2000," he said. "*That* was the first monkey paper." Indeed, writing in an Institute of Electrical and Electronics Engineers journal in June 2000 (five months before the Nicolelis paper), Schwartz and his colleagues described how they used penetrating electrodes to read motor neurons in real time via a computer. Schwartz and his colleagues did not link the animal's brain to a robotic arm, but they indicated it was a feasible next step, writing, "This architecture . . . will eventually be used to drive the robotic arm."

Fifteen years later, the sting that Nicolelis drove an arm first remains palpable. Schwartz had been working neck and neck with Nicolelis. The difference was that he hadn't been in a rush to publish, and the Duke team beat him to the punch. "That really pissed me off. It was shitty control—even in 2000," Schwartz said. "We had 3-D robot control about three years before Nicolelis came out with his paper. We didn't think to publish it. It was like, okay, we got this working, right? We had the robot in one room, the monkey in another room . . . We didn't make a study of it. We just thought, isn't that cool?"

Schwartz was determined not to repeat the mistake, publishing his 2002 study in the pages of *Science*, where no one could miss it. The experimental paradigm was similar to work he'd done with the Georgopoulos lab—a "center-out" task where monkeys had to move a central cursor to one of eight targets.

By then, both Nicolelis and Donoghue had decoded neural firing patterns to re-create arm movement. In these so-called open-loop paradigms, monkeys simply performed a physical action—reaching, say, or moving a joystick. Researchers simultaneously reconstructed the movement from brain activity, but the animals were none the wiser: they simply completed their tasks and received their juice rewards. In Nicolelis's work with Belle, for instance, the owl monkey wasn't aware of the arm in the next room, and that is to say nothing of the arm in Cambridge. All Belle had to do was control the joystick. The computer did the rest.

What set Schwartz's 2002 study apart was that he "closed the loop," giving the animals visual feedback and direct cortical control of the cursor. Researchers began each day by taking "baseline" recordings, using their electrodes to chart the tuning properties of around twenty neurons as the monkey moved the joystick to the left, to the right, up, or down. Meanwhile, Schwartz used a shield to block the animal from seeing its arm. This focused the monkey's attention on the monitor as researchers built a "decoder," an algorithmic model that determined the directional tuning properties of individual neurons as they relate to arm movement.

With the decoder in place, the experimental paradigm remained largely familiar. The shield still blocked the monkey's arms. The computer screen still presented a cursor and eight targets, and the juice tube remained near the animal's mouth. But Schwartz and his colleagues introduced two key changes: restraining the monkey's arms and removing the joystick. With the joystick offline, the researchers shifted the cursor's control mechanism, placing it under direct command of the animal's brain.

They had closed the loop.

The monkey knew exactly what it needed to do physically to earn its reward, but with its arms restrained, the animal had to discover a new way to move the cursor. During the first few days, the monkeys struggled against the arm restraints. As they did so, the cursor began to move, and within a few days each monkey realized it didn't need to move it limbs at all. Thinking was enough.

The monkeys had immediate visual feedback. As the implanted electrodes ferried information to the computer's algorithm, the algorithm transformed the patterns into control commands for the cursor. It happened in real time, as responsive as a joystick, which allowed the animals to correct whenever the cursor overshot the mark.

And this was one of the study's great findings. Previous work had assumed that neurons would always fire with the same intensity for a given movement, more so for certain directions, less so for others. But what Schwartz and his colleagues found was that once the animal realized it had neural control of the cursor, it was able to adjust its brain signals for better control. Rather than being fixed, tuning properties were malleable. Neurons could adjust their behavior to better interact with the new interface.

Said differently, Schwartz and his colleagues were watching as the animal—the animal's brain—learned how better to control the cursor. They were watching individual cells transform their behavior. They were observing learning at the cellular level. "When learning takes place, what does that mean? That means that the relationship between a neuron and whatever thing you're trying to learn, that relationship has changed. That is learning," said Schwartz, describing how individual cells changed their behavior to better control the cursor. "We can actually say the amount of learning is proportional to how much that tuning function changes."

The monkeys became more adept with the cursor as the study progressed. Perhaps that's not too surprising. After all, coordination, be it throwing a ball or controlling a computer, takes practice. But this matter-of-fact reading doesn't take into account what Schwartz observed at the cellular level, where neurons progressively shifted their preferred firing direction to better interact with the electrodes.

The neurons were not merely learning a new skill. They were transforming their biological patterns to adapt to a nonbiological interface. Importantly, the transformation wasn't permanent. As the study progressed, neurons began to show individual behaviors for distinct tasks and modalities. The cells fired differently when a monkey used its brain to move the cursor than they did when the animal moved its arm in the same direction.

What this meant, conceivably, was that while the algorithm

had initially enabled the animal to mimic hand movements to control the cursor, the brain eventually embraced *the cursor itself* as an entirely new appendage—an appendage that was unlike its biological counterparts and required brain function all its own. The brain was not translating arm movements to control the cursor. Rather, it had reorganized to embody the cursor itself. Perhaps most intriguingly, it didn't take a few weeks for the brain to adapt. It took minutes.

The suite of experiments cemented Schwartz's reputation as one of the field's top monkey men. Nevertheless, when it came time to submit a proposal for the Revolutionizing Prosthetics program, Schwartz wasn't brought on board. The agency awarded him a separate grant instead. It wanted him to keep working in monkeys.

4. BAD CODE

"I think it says 'cough,'" Brookman said, sitting up in bed at the epilepsy-monitoring unit at Barnes-Jewish Hospital in St. Louis. It had been a few days since Eric Leuthardt implanted his electrode grid, and after suffering rolling seizures, Brookman finally seemed more lucid. David Bundy, Nick Szrama, and their fellow graduate students had already cut several sessions short when Brookman fell into a seizure or was too dazed to understand the task. But Brookman had forgone his pain medication today. He was sitting up in bed, and his voice, though still subdued, was stronger. He folded his arms over his bare torso and peered at a monitor as he spoke into a microphone.

"It should be the same words, just pick one and take a guess," Szrama said, standing to the side of the bed while monitoring a laptop. "She's going to say either 'ka,' 'ga,' 'da,' 'ta,' or 'pa.'" Szrama wanted Brookman to perform the McGurk task, so named for its inventor Harry McGurk, who in the mid-1970s noticed that when we receive conflicting auditory and visual stimuli, the visual information will often override the auditory stimulus.

When a subject sees and hears a person say a phoneme like "ba," for instance, he will correctly perceive to have heard and seen the phoneme. But when researchers keep the auditory stimulus constant ("ba") and change the visual stimulus to a different phoneme, such as "fa," the subject will perceive he's heard what his eyes tell him, incorrectly perceiving that he's heard the phoneme "fa." The illusion disappears instantly when the visual stimulus is removed. When the subject closes his eyes, for instance, he will correctly perceive the auditory input "ba."

We like to think that our senses are accurate—that our eyes and ears perceive stimuli as they actually occur. We like to think that no matter what passes before our visual field, our ears will accurately perceive incoming sounds and words. But the McGurk effect shows that that's far from the case. Our visual and auditory senses are deeply interwoven. They are not merely dependent on each other; they actively influence each other. More troubling yet, when those senses receive contradictory information, the brain will incorrectly perceive, or override, stimulation in its effort to make sense of the world.

Neuroscientists can observe the McGurk effect and describe the phenomenon, but very little is known of its underlying neural mechanics. How, exactly, does the brain synthesize incoming auditory and visual information? What neural mechanisms allow vision to take precedence over hearing? And why, even when we know about the McGurk effect, do our brains continue to perceive sounds incorrectly when they conflict with visual stimuli?

Using the electrodes Leuthardt had implanted in Brookman's brain, Szrama and his colleagues were trying to get a better sense of just how these two sensory systems interact and influence each other. For years, Leuthardt had been trying to lay the foundations of a speech prosthetic—a BCI that could read the neural correlates for language and translate them to a system for the mute, or perhaps for a soldier who needs to communicate silently. His lab was trying to gain insight into how the brain breaks down words by working with individual phonemes, the monosyllabic building blocks of language. Of course, words are more than a series of sounds. They have associative meanings that are tied to experience and ideas. And while words and language are undeniably composed of phonemes, it's not clear that the brain conceives language primarily in these component parts, building words sound by sound. It could be that the brain's creation of language is more meaning based, forming distinct neural patterns depending on the context of a given word or a shade of meaning.

Viewed in terms of motor function and auditory input, however, phonemes weren't a bad place to start. Dressed in slouchy khakis with his tie slightly askew, Szrama asked Brookman to don a pair of headphones and repeat the series of phonemes—ba, da, ka—he heard. "It seems like they're all the same," Brookman said after a few moments. It took him a minute to understand the task, but once Brookman got the hang of it, Szrama increased the task's difficulty, asking him to watch a video of a woman silently mouthing the phonemes.

The idea was first to record Brookman's brain activity as he simply heard the phonemes. Then researchers would record as Brookman watched the woman mouth the sounds with no audio. Once the researchers had recorded his neural response to these separate stimuli, they would perform a third recording, this time combining the correct audio and visual feeds. The final step would be to induce the McGurk effect, hoping to shed light on how the brain gives priority to visual stimuli when faced with contradictory sensory input.

"Okay, this time you're not going to hear anything; you're just going to see a woman speaking," Szrama said. "You're going to have to try to guess the words she's saying."

"Oh my God," Brookman said as the woman with short brown hair and a blue turtleneck appeared silently on the screen.

"Just try to guess and say what word she said," Szrama said.

Brookman hesitated.

"Just try your best," said Szrama.

"Oh my God."

"Can you take a guess?" he asked. "How are we doing?"

Not well. Brookman was confused. He was also starting to fade. His foot twitched under the hospital sheets, and his eyes kept falling shut.

"Try to keep your eyes open," Szrama said, his eyes trained on a computer monitor. But it was too late. Brookman couldn't keep his eyes open for more than a few phonemes. He would guess at every third sound, but soon enough his glazed eyes fell shut, rendering vast swaths of data unusable.

"Can't do it?" Szrama asked.

•

Brookman's mother had kept off to the side throughout the testing. A woman in her sixties, she worked as a forklift operator in southwestern Missouri, where she lived with Brookman's stepfather. She wore her straight brown hair in a ponytail with bangs. She had traveled with her sister to St. Louis for the surgery, and they had remained at Brookman's bedside throughout his stay. But whereas Brookman's aunt spent those days encouraging him, praying for him, reminding him of life outside, and asking if there was anything he needed, his mother did her best to stay out of Brookman's line of sight. "It just works better that way," she said.

Three years after his initial surgery, Brookman's relationship with his immediate family had deteriorated. His seizures made it impossible for him to safely drive a car or hold a steady job, but he'd nevertheless decided to move two hours away from his hometown of Joplin, Missouri, where his mother and two siblings lived.

It hadn't always been that way. As a child, Brookman had been close to his mother, a "mama's boy" and a "pleaser," as she put it. Nevertheless, his neurological troubles started early. He suffered his first seizure when he was only eighteen months old. He'd been playing in the kitchen, when he suddenly fell over, his eyes rolling back in his head. "He was like a limp rag doll," said his mother, who rushed to pick him up. "I thought he was dead." She ran to a neighbor's house to call an ambulance, but Brookman recovered almost immediately. "By the time they got there, he was already up like nothing ever happened."

In the weeks that followed, Brookman's mother took him to see a neurologist, who diagnosed him with a seizure disorder. The family tried various drug regimens over the years, once consulting an herbalist who persuaded them to take Brookman off his medications all together. "He had eleven seizures that night," Brookman's mother said. "That's when I knew he was going to have to have his medication forever."

Brookman's seizures came regularly when he was a child, but they took a more sinister turn around his eighth birthday, often causing him to lose consciousness. Convulsions began to plague him at night, and his mother could hear him moaning in his bedroom as she watched television. "I'd look, and he was basically thrashing around in the bed," she said. "I watched just to make sure he was okay."

Meanwhile, Brookman was suffering other disorders as well. Doctors diagnosed him with Tourette's syndrome when he was seven years old, and by fourteen he'd been diagnosed with bipolar disorder. Brookman was already on a heavy drug regimen, however. His doctors feared he'd have an adverse reaction, so both conditions went untreated.

Brookman's Tourette's syndrome caused him to twitch violently. He could control it during the day, but the spasms and seizures took hold at night. His bipolar disorder, by contrast, was uncontrollable. He flew into rages, and as his conditions intensified, his relationship with his family became strained. "He blames me for the way he was born," said his mother. "He looks at me with this glare, like it's all my fault."

Perhaps most devastating, though, was the shame. Brookman began to wall off his life with secrets, hiding his conditions from friends and women he wanted to date. But with the constant threat of seizures, mood swings, and twitching fits, there was only so much he could hide. He was increasingly alienated from his family, and at thirty years old he'd never been in a romantic relationship. "I would never put a girl through what I go through every night," he said. "I'm so embarrassed by what I have that I will not get into a relationship." The same held true for friends. "He doesn't want me to meet any of them," his mother said. "I don't know if he's afraid that I'm going to tell them something or what, but I've never met them."

Living in rural Missouri, Brookman learned that he might be a candidate for surgery only in 2007. He'd read a few articles, and he asked his local neurologist if he might be eligible. "I wanted to have a day and a night without having a seizure—without having it on my mind every day," he said. "I wanted to feel like an actual individual that doesn't have disabilities. I wanted to feel like they do."

Brookman seemed cured immediately following his first surgery in 2008, going a day without seizures. But the price was steep. Simple math was a mystery when he emerged from the operation. He couldn't read or write. He had lost command of his ABCs, and he couldn't remember his mother's name. "I was like a new baby being born," he recalled.

But Brookman's brain was already forming new synaptic connections to perform these old skills. His brain worked quickly to compensate for the areas that had been removed, and within a few days his mother's name had returned to memory. He was reading and writing within a month, and he learned to walk again by following his aunt through the rehabilitation center.

But Brookman's brain didn't stop there. Neuroplasticity has a dark side, and just as Brookman's brain forged new pathways to replace those knocked out by the surgery, so too did it begin to fashion new epileptic trails. "Oftentimes the brain wants to have seizures, meaning that the brain is plastic," said Leuthardt. "You can have positive plasticity, meaning the brain re-accommodates its function, but it can also re-accommodate dysfunction, meaning that if your brain is trained to have seizures, it may resume that ability." Brookman's seizures soon returned, eventually resuming at their former intensity.

One thing that did not return, however, was Brookman's long-term memory. Today he says he remembers little before the age of seventeen. The surgery also affected his short-term memory, and Brookman now relies on copious Post-it notes and reminders that he thumbs into his cell phone.

He was calmer when he returned home in 2008. He didn't seem to remember the difficulties he'd had with his family, which eased tensions around the house. "It might be a good thing to have a little memory loss," his mother said.

When a friend moved two hours away to Tulsa, Brookman decided to move, too. Over the next few years, he tried to reinvent himself, taking care not to overheat playing basketball and telling all who asked that he worked only occasionally because his family had money. Still, there were times when he couldn't hide his illness, such as when he had a seizure while working at a clothing store or when his friend became suspicious of the groaning that emanated nightly from his bedroom.

" 'I hear you every night screaming like somebody's killing you,' " he remembered his friend telling him when they finally discussed it. "I'm just so sick and tired of dealing with it," Brookman said. "I'm a very positive, very laid-back individual, but with what I have, I don't even care to live."

As it stood, a second surgery was Brookman's only hope. Like his earlier procedure, the operation would carry a 50 percent chance of success. It also carried all the previous risks like memory loss, an inability to write or work with numbers, blindness, and even death. But Brookman was out of options, and three years after his first surgery he was willing to gamble.

"You say why, why this child? My other kids have children. They're married and have full lives, but he has nothing. He's so lonely," said his mother. "He'll call saying he wants to be dead. Then I'll start crying, saying it's going to get better. But what are you going to say? That he's not going to get better and push him over the edge? It's something that's always at the back of my mind."

•

Epilepsy remains one of the brain's more mysterious diseases. Unlike neurological disorders that involve cell deterioration or the formation of plaques, epilepsy uses the brain's own circuits. Just as physical actions are produced by neurons that communicate via synapses and neurotransmitters, so too is a seizure produced by neural communication. The difference being that while normal brain function involves neurons firing at different rates as they volley information back and forth, a seizure is marked by large-scale hyperactivity, as whole groups of neurons fire in synchronized blasts.

In that respect, epilepsy is not unlike a computer virus—bad code that hijacks the system's circuitry to run its malicious program. Perhaps more provocatively, epilepsy's use and creation of neural pathways is in some ways indistinguishable from normal brain function. "Epilepsy reflects the way the brain is organized and the way things are connected normally," said Brookman's neurologist, Edward Hogan. "Epilepsy functions like other brain functions. We define it as abnormal, but that's because most people don't have epilepsy."

Like other brain functions, epilepsy doesn't involve just one neuron or one group of neurons. It moves across neural networks, recruiting otherwise normally functioning groups of cells. "Groups of neurons work together to generate lots of things. In epilepsy's case, it's pathological, but everything we do every day relies on big parts of the brain working together," Hogan said. "Our hope is to remove enough of [the epileptic focus] region that we knock out the network. If we're able to do that, then we're able to stop the process that causes seizures."

For Brookman, that region lay deep within his parietal lobe. Like the rest of the brain, the parietal lobe is wrapped in neocortex—the furrowed mantle of recently evolved gray matter that sheathes the organ's white matter. The neocortex's expansive wrinkles both enfold and partially constitute the brain's dual hemispheres, wrapping them like an orange peel. Brookman's seizure focus resided on an interior fold of gray matter—so deep, in fact, that it faced the opposing hemisphere.

Leuthardt might simply have trained his scalpel through the central valley that divides the brain's hemispheres, taking a direct route to the epileptic seat. But the source of Brookman's seizures lived deep in this valley. It hid below a rich matrix of veins that bridge the divide, and cutting through the venous network would mean torrential hemorrhaging. Instead, Leuthardt would need to cut through Brookman's healthy brain tissue, digging a proverbial hole to China—through gray matter, white matter, and back to gray matter—as he aimed to remove the specter that had terrorized Brookman since he was an infant.

"Getting there is very hard to do," said Leuthardt. "You have to go straight through the brain." Charting a surgical course through such a vast swath of neural space comes with significant risks. The parietal lobe is home to the somatosensory cortex, a brain region essential to processing the body's senses. The optic stream courses through the lobe en route to the visual cortex. Similarly, the lobe is foundational to verbal memory, language, and a host of other functions. Cutting directly into the lobe, Leuthardt risked removing or damaging some of these areas and leaving Brookman with even more disabilities than he already had.

Still, the surgeon was confident he could remove the seizure focus without unduly injuring his patient. His path would have to be precise—the surgical equivalent of navigating between Scylla and Charybdis as he avoided areas devoted to synthesizing sensory information, the optic stream, and regions associated with memory and language.

So it was that a few days before the surgery, Hogan brought his own system into Brookman's hospital room. The hospital staff had removed the leather straps they'd used to keep him from tearing the electrodes from his head, replacing them with a pair of thin ropes. Not that they were needed. Brookman was slipping in and out of sleep as his mother and aunt looked on from their pair of recliners.

Dressed in a white lab coat, Hogan sat at a portable table peering at a computer as he studied Brookman's brain waves. He moved the computer's pointer to a window titled "Cortical Stim" and pressed the "start" icon. As the system delivered a small pulse of electricity to a pair of electrodes, Brookman groaned and the right side of his face began to contract. Hogan drily noted his patient's reaction before moving on to the next pair of sensors on the grid. He delivered another charge. Brookman groaned again as his head turned involuntarily to the right. It was like watching a puppet move. When Hogan stimulated another pair of electrodes, a confused look came over Brookman's face as his mouth began to move. "I'm going to be heading down toward the leg," Hogan said as he engaged yet another set of electrodes, causing Brookman's right arm to rise.

Specific areas in the body correspond to particular brain regions, and Hogan was using the electrode grid to create a functional map of those areas, determining which regions were associated with which part of the body. The map would give Leuthardt a better idea of how best to chart his surgical course.

Still, certain key functions could not be avoided. Although the visual cortex resides in the occipital lobe in the rear of the brain, many of the tracks that lead from the eyes move through the parietal lobe, where Leuthardt would be working. "This is just anatomically in the brain," Hogan said. "If you knock out those tracts, you'll lose the pathways that carry visual information."

As Hogan continued across Brookman's cortex, there was a large column—two rows of electrodes wide—that elicited no physical response to the system's pulses. "Seems safe with motor," an assistant said, keeping track of Hogan's progress. "Can you feel anything?" Hogan asked. Brookman moaned no, but it was unclear that he'd even understood the question. "My concern is that he was asleep," Hogan said after completing the map. "If he's awake and we invoked anything that would've caused a sensory change, you'll see him kind of wake up, even if he can't describe it. But if he was asleep, he might get a feeling and just sleep through it."

Nevertheless, between the brain mapping and the seizure monitoring, the doctors had formed a detailed portrait of Brookman's brain. They not only had localized the seizure focus but also could steer clear of critical areas. "The first time it was, you know, it's a big brain," said Hogan. "We just had a few electrodes over the area that caused seizures, and if you just have one or two electrodes there, it's hard to map."

This time, by contrast, Leuthardt had placed some one hundred electrodes over the seizure focus. He knew exactly where to go, and although the surgery would almost certainly leave Brookman partially blind, they deemed it a sacrifice worth making.

"He's always been very, very strong in favor of doing the surgery, but there are a lot of variables. We have to feel that it's the best thing for him," said Hogan. "If someone has any doubts about it, we just don't move forward. It's not worth it."

•

In the days that followed, Szrama, Bundy, and the rest of Leuthardt's research lab donned ties, tucked in their shirts, and headed over to the epilepsy-monitoring unit. Brookman, however, kept slipping in and out of lucidity. One minute he was warring with hospital staff; the next he was trying to charm the researchers and neurologists. But between the drugs, the seizures, and his impending surgery, Brookman was often too dazed to understand even the most basic tasks.

"If the words are unrelated, just click the mouse button. If they're related, don't do anything," Szrama told him, explaining a new research task.

"Oh, so it's more than one letter?" Brookman asked.

"There'll be words like 'autumn' and 'fall.' Those words are related, so you don't need to click anything," Szrama said, instructing him to click the mouse only when he heard unrelated words.

The researchers were hoping to "prime" Brookman's brain,

getting him to associate specific meanings with homonyms. For example, they hoped the word "autumn" would prompt him to associate the word "fall" with the season as opposed to tumbling to the ground. The idea was that once they'd primed the brain for a homonym's specific meaning, they could study the brain's neural response to what (at least physically) was the same auditory stimulus ("fall" and "fall"), enabling them to see how the brain represented meaning in language.

Earlier fMRI studies had shown that depending on their meaning, homonyms could be associated with different areas of the brain. But fMRI only charts blood flow, taking a series of snapshots of how a word is represented spatially in the brain. Leuthardt's electrodes, by contrast, could give them what's known as temporal resolution, monitoring how continuously shifting brain waves reflect different meanings.

At least that was the idea.

"So if this 'queen'—'king,' does that mean that I need to, uh," Brookman replied, watching the prompts flash across the screen. Another pair of words quickly replaced them. "See 'arrest' and 'watch' isn't," he said, trying to keep up as the new pair flashed and was again replaced, "so if it's a 'cat,' and uh . . ."

It was one of the pitfalls of Leuthardt's research model. Unlike his colleagues who worked with animals, Leuthardt worked exclusively with intraoperative human patients. Working with humans has its benefits. (First and foremost, they can tell you what they're thinking.) But it also has its limitations. Patients like Brookman are under extraordinary stress, and the ordeal can incapacitate them to the point that they cannot fully participate in the research.

"He is having a lot of seizures that are knocking him back, but that's not uncommon," Leuthardt said over lunch one day between surgeries. Brookman's rolling seizures meant he could only partially participate in the research, but they had also given Leuthardt a lode of information about his seizure focus. The electrodes had pointed with near certainty to the origin of Brookman's epilepsy, and his doctors decided to move the surgery up by several days.

"We're going to favor being aggressive. But exactly how aggressive, and what are my limits? I don't know yet," Leuthardt said as he prepared for the surgery. "Maybe we did not do an aggressive enough surgery the first time around. There's a real tug-of-war between being surgically aggressive and taking as much as you can versus being more conservative and lessening the deficit. You're always walking this tightrope."

•

Early the next morning, Leuthardt briskly clipped the stitches from Brookman's scalp. Blue, pink, and red electrode tails sprouted like a Mohawk down the center of Brookman's head as Leuthardt worked quickly to peel back the scalp. Removing the piece of skull he'd fastened with temporary titanium plates, the surgeon placed the bone flap in a nutrient-rich solution, revealing the blood-caked globe of dura mater beneath. The clear grid of electrodes began to emerge as Leuthardt used a probe and suction to clean the area. Five depth electrodes studded the grid, whose wires tunneled under Brookman's scalp and exited a few inches from the incision Leuthardt had made earlier that week.

"Pull the ribbon," he told a nurse after clipping the grid's wires to untether the electrodes. Giving the wires a sharp downward tug, the nurse yanked the tails from the exit port.

Leuthardt then used a pair of micro-scissors to cut a corner of the grid. Trading out the scissors for a pair of tweezers, he lifted the grid from Brookman's brain, revealing its faint impression on the tissue. Once Leuthardt had irrigated the brain's surface, a surgical resident used a metal depressor to expose the cavity from Brookman's earlier surgery. "Our culprit is right down there," Leuthardt said as he probed the cavity's sides.

A song by the English band the Cure played over the room's

sound system as the surgeon methodically cut a horseshoe-like arc into the surface of the brain. Using the cauterizing scalpel, he described the circumference of the tunnel he would create to reach the epileptic seat. As he burrowed deeper into the brain, Leuthardt periodically switched to a pair of micro-scissors to cut through blood vessels, cauterizing the wound and leaving blood-absorbing cotton strips in his wake.

Computer monitors showed where and how deep he should cut. Leuthardt would occasionally consult these digital maps, but he hewed mainly to the path formed by one of the depth electrodes, which joined a second electrode he'd implanted at the disease's epicenter.

By then, Leuthardt had adopted a curt surgical patois of one-word commands: "bipolar," "micro-scissors," "probe."

"There it is," he said as his cylindrical incision finally reached a second electrode. "We know the seizure focus is coming from between these two electrodes."

Using a pair of cup forceps, the surgeon pulled back the resection area—a core of brain roughly the size of a wine cork—to expose the bad brain. He pulled firmly to remove the entire cylinder, examining its flan-like consistency before dropping it into a bottle bound for the pathology department.

Turning back to the cavity, Leuthardt then began to cut deeper still. He was now at the heart of Brookman's epilepsy. He no longer consulted the digital map as he assessed the wedge of fibrous tissue that had tortured Brookman for thirty years.

"Weird," he said as he removed the rogue tissue, which was about the size of a mandarin orange section. It was significantly darker than the rest of the brain, rubbery and fibrotic.

With the traitorous piece of brain removed, Leuthardt used a microscope to move quickly up the cavity wall, pulverizing and suctioning odd-looking bits of brain while cauterizing any areas that were still bleeding. Once the bleeding was contained, he removed the remaining depth electrodes before reattaching the dura mater, draping it over the resection cavity and suturing it back into place. He then moved quickly to close the surgical site, anchoring the skull flap with titanium plates, smoothing the scalp, and closing the incision with hundreds of staples.

"He's waking up," Leuthardt said as he hurried to wrap Brookman's head in surgical gauze. "Let's go."

5. SCREW THE RATS!

Leuthardt had known BCIs could potentially restore movement, but it wasn't until he witnessed the neurosurgeon Jeffrey Ojemann use electrodes to map the brain of a surgical patient that he realized it could unlock some of the brain's more basic mysteries. Using an electrode grid he'd placed on the patient's brain, Ojemann drove small electrical currents through the sensors as he asked the patient, who was conscious, to think about rotating an object. The surgeon was checking for cognitive errors. Each time the patient faltered, Ojemann knew that the stimulating electrode had interrupted a critical brain area beneath—a region to be avoided during surgery.

All in a day's work for many neurosurgeons, but for Leuthardt, then a surgical fellow already dreaming of a neuroprosthetic future, it was a revelation. He realized that by placing electrodes on the surface of the brain, you could not only eavesdrop on the cortex but also directly interact with it, introducing information to the neural matrix. "Dear Jesus!" he thought as he watched his mentor engage the patient's brain. "That is a great model for science."

Leuthardt was a mere twenty-eight years old. He'd studied theology before heading to the University of Pennsylvania for medical school and a residency in St. Louis. Unlike many of his colleagues, however, Leuthardt hadn't always been an overachiever. He had been a fighter growing up in Cincinnati, where his mother raised him alone after his father left. Though born in Boston, Leuthardt spent his early childhood in Europe. Once in Cincinnati, however, he was a bookish European kid, the son of a divorced mother at an all-white Catholic school in Middle America. He was interested in science and the future, but he struggled in school, both socially and with his teachers. "I got really used to having it be very unpleasant," he said. "I got picked on tremendously." Leuthardt didn't play sports, and he kept close to the few friends he had. He learned to be self-reliant. He also learned to protect himself, often getting in fights and being punished with detention.

His summers turned glorious, though, when he started volunteering in the research lab of Keith Crutcher, a neurosurgeon at the University of Cincinnati. Crutcher was a religious man, but he was also a scientist who was generous with Leuthardt, affording him a glimpse of a much larger world than the one he inhabited. "He was a guy who'd masterfully integrated this analytical reason with this deep, deep sense of faith," Leuthardt said. He ended up volunteering for four summers in Crutcher's lab, where he spent hours peering into a microscope dissecting nerve ganglia in chicken embryos. It was delicate work, and Leuthardt's native skill with his hands caught the attention of Wayne Villanueva, a neurosurgery resident who was spending a research year in the lab. The young doctor invited Leuthardt to observe a brain surgery.

"That was a deep awakening experience for me. There was something so fundamental to it—the intensity—it was messy and deeply meaningful," he said. "That really set my trajectory."

Leuthardt went on to study theology and biology as an undergraduate at Saint Louis University. He'd never been rich. His mother had raised him on her art teacher's salary, and when she lost her job, she'd had to move in with his grandmother. He was still hounded by some of his childhood anxieties, but he pushed to reinvent himself through education and accomplishment. "There was nothing to fall back on," he said. "I became a very determined workaholic not to get pushed around by the world—you know, never again. The world would operate by the force of my will, and medicine, at least from what I was exposed to, was the most stable and prestigious."

He went on to medical school at the University of Pennsylvania, later completing a neurosurgery residency at Washington University in St. Louis and a fellowship at the University of Washington in Seattle. Leuthardt had always been fascinated by the brain and mind—how this three-pound organ defines us as humans. We are it. It is we. Yet in many ways, we know nothing about it. Theologians, artists, writers, and philosophers have tried to describe it with terms like ego and soul. But as Leuthardt watched Ojemann communicate directly with the brain itself, he realized that unlike the metaphors of the past, neuroscience, neurosurgery, and BCI promised a biological understanding of consciousness and cognition. "It was this true interaction with the substance of a human being," he said. "That's when I became really interested in epilepsy. You're directly interrogating the human brain. A lot of surgery is pick-and-ax type of stuff—sew the vessel together, or pull this out—but this was a way to get at science-y, higher cognitive type of stuff."

Andrew Schwartz, John Donoghue, and their cohort had been publishing results for a few years by then, and it was becoming clear that BCIs would soon be able to conduct a two-way conversation with the brain—not the person, per se, but rather the mysterious neural architecture that undergirds consciousness and personality. "It became very clear to me that this was the future," he said. "I already had these philosophical interests, but once I saw this convergence by which you could open up a whole other frontier of the human experience, I didn't hesitate. I jumped on it."

•

The field's potential for basic science appealed to him, but Leuthardt was equally captivated by its promise to transform not merely the medical field but the very experience of being human. In essence, Leuthardt saw in neuroprosthetics something unprecedented in roughly 200,000 years of human history: an entirely new pathway for the brain to manifest its intentions. By wiring the brain directly to a computer, researchers might unyoke thought from its traditional confinement to the brain and body, bypassing a biological dictate that has held sway since the first invertebrates formed loose nerve networks to move about.

It seemed like an inviolate rule of biological evolution: organisms would rely on a class of interconnected, information-carrying cells known as neurons to direct muscles, their sole means of interacting with the environment.

But with monkeys taking mental control of cursors and rats moving feeding levers, these early researchers were tilting at a radical new vision of how the brain might interact with the physical world. By grafting the brain directly onto a computer, they were clearing the way to extend the body's biological nervous system of dendrites and axons into a digital realm of 0s and 1s.

As Leuthardt saw it, neurally controlled appendages were just the beginning. The increasing digitization of the environment meant the brain could someday interact wirelessly with a host of surrounding technologies—everything from smartphones and laptops to climate controls and lighting. "Maybe for no other reason than I think it's amazingly cool," he said, "but that's cracking something." After all, we only need to move a mouse in three directions (up-down, side to side, and click) to gain control of a computer. That's merely three degrees of freedom, a feat some researchers had already demonstrated. Why couldn't a neural augment communicate with a smartphone or wearable tech, granting users access not only to the Internet but also to a communication hub for surrounding technologies?

It was a thrilling prospect, and one that opened a broad challenge to time-honored notions of evolution, biological order, and necessity.

Our best estimates place the age of the world at roughly 4.5 billion years old. The fossil record indicates that simple one-celled organisms, the ancestors of today's bacteria, ruled the earth until about 1 billion years ago, when these life-forms began to mutate, ushering in a generation of multicellular organisms not unlike today's sponges. The first animals crawled from the sea around 500 million years later, and it took another 500 million years for hominids to arise, which occurred a mere 7 million years ago.

Through it all, genes ruled. Counterintuitive though it may be, a gene's objective is not the survival of a particular animal or species. A gene's primary aim is its own replication. The survival of the animal it happens to construct—or, as the evolutionary biologist Richard Dawkins would have it, the genetic "vehicle"—is incidental, important only insomuch as it provides a successful means for genes to survive and replicate. It's a distinction that is easily overlooked, but in genetic terms, an animal is little more than a temporary structure genes build to ensure their replication.

In this sense, it's largely irrelevant whether a gene survives in a flatworm, a giraffe, or a human. What matters is that collections of genes combine to create biological vehicles—organisms—that are successful in specific habitats.

Of course, genes optimize a biological vehicle through the fumbling process of genetic mutation, the evolutionary mechanism that has governed life for the past 3.5 billion years. At least it was until 160,000–200,000 years ago, when our species arose from the hominid line to offer what some evolutionary biologists describe as an alternative to the genetic monopoly: culture.

Just as a biological vehicle protects the genes it houses, so too does culture allow groups of humans to survive by forming what the evolutionary biologist Mark Pagel calls "cultural survival vehicles." Pagel argues that, like the protective plates of an armadillo, or the kinetic muscle-and-bone architecture of a cheetah, culture enmeshes us. It empowers groups of humans with shared language, common identity, and an inherited knowledge base.

It is the crowning adaptation of our species and wraps us in a protective mantle that defines us, enabling our survival. In that sense, culture is not so different from a beehive or an ant colony: a collective activity that functions almost as an organism itself, protecting individual members that would quickly perish on their own.

Like social insects, humans exist almost exclusively in organized groups. We may not be able to draw physical boundaries around a culture, but our cultures, no less than a body for genes or a hive for bees, have been critical for our success. The difference, of course, is that while social insects have defined activities, individual members of a culture are relatively free to play a host of roles. What's more, culture is cumulative, meaning that ideas—the tempering of steel, say, or the invention of penicillin—can be transmitted not only from one person to another but also from one generation to the next. As Pagel writes in *Wired for Culture*, it is hard to overemphasize its evolutionary importance:

> Our invention of culture . . . created an entirely new sphere of evolving entities. Humans acquired the ability to learn from others, and to copy, imitate and improve upon their actions. This meant that elements of culture themselves—ideas, languages, beliefs, songs, art, technologies—could act like genes, capable of being transmitted to others and reproduced. But unlike genes, these elements of culture could jump directly from one mind to another, shortcutting the normal genetic routes of transmission. And so our cultures came to define a second great system of inheritance, able to transmit knowledge down the generations. For humans, then, a shared culture granted its members access to a vast store of information, technologies, wisdom, and good luck . . . Having culture is why we watch 3D television and build soaring cathedrals while

> our close genetic relatives the chimpanzees sit in the forest as they have for millions of years cracking the same old nuts with the same old stones.

Pagel argues that our genes themselves have evolved—not to inhabit any one physical habitat, but rather to inhabit the most important environment for our species: the social environment. Our genes have produced a neural architecture that creates the mind, a mind that is flexible enough to embrace any cultural environment it happens to be born into.

This behavioral flexibility enables us to acquire what Pagel calls social learning. Whereas other animals can learn new behaviors, humans can not only learn a new behavior but also understand the utility of what we've learned. We can translate it into a new environment, passing novel behaviors on to others and even improving on them.

It's the reason we now inhabit all corners of the globe. While other species rely on genetic adaptation to exploit new environments, humans need only tap our vast cultural resources to take on novel traits.

In earlier times, this might have meant learning that fletching makes a spear fly straight or that iron tempered with carbon forms a stronger metal. More recently, antibiotics have enabled us to ward off once-deadly bacteria. Advances in genetics have created new breeds of disease-resistant crops, and computers have enabled us to collect and analyze unthinkably large troves of data. These are just a few of the millions of ideas that have moved through the human population, enabling us not only to take on adaptive traits with lightning-like speed but also to thrive in environments we once found hostile. It might have taken the modern polar bear eons to develop the layers of fur and fat that enable it to live in arctic climates. By contrast, we can adopt the traits necessary for arctic living by tapping a veritable genome of ideas—everything from leather tanning and high-tech fabrics to extracting and refining fossil fuels and internal combustion engines.

"Instead of adapting to the demands of any one physical environment, our genes have evolved to use the new social environment of human society to further their survival and reproduction," writes Pagel. "These are the adaptations that have wired our minds and bodies for culture."

Through culture, we have in some essential way transcended nature. We have become the drivers of our own evolution, and though we may not foresee the cumulative effects of certain ideas, we are nevertheless able to introduce new ideas or mutations into our cultural survival vehicles—our hives. "Ideas in brains are the new information carriers that are shaping evolution," said Leuthardt. "There's a whole new organism that is created that we're not aware of. Humans and the cultural and social infrastructure are all part of this organism. It's self-maintained, and, I would argue, self-aware beyond our potential capacity to understand its awareness. Call it *Homo socialis*, and just as the biological nervous system becomes more developed, with neuroprosthetics you're seeing a new nervous system develop . . . You create a much more integrated organism. Just as the printing press created a more integrated organism, just as telephones created a more integrated organism, just as the Web, Twitter—all of these things."

Culture might have freed us from some of evolution's genetic fatalism, but there is no disputing that in many ways we remain beholden to our biology. We still get old, and barring birth defects, we are each born with a pair of arms and legs. No matter our cultural achievements, we have historically remained bound by genetic evolution and our neuromuscular mode of acting in the world. But it was precisely this evolutionary tenet that early BCI researchers were threatening to disrupt. This was not a trait that had been handed down through the generations. Rather, by splicing a digital nervous system onto the brain, they were promising to bypass the slow, hit-or-miss modulations of genetic evolution, swapping it out with a deliberate, nonbiological innovation.

"It's an engaging notion," said Leuthardt. "Your ability to interact with the world is no longer constrained by the length of your arms."

•

The image of the cyborg has inhabited a contentious plot of cultural real estate ever since Manfred Clynes and Nathan Kline first coined the term in their 1960 paper audaciously titled "Cyborgs and Space." The Soviets had humbled the Americans three years earlier by launching the unmanned *Sputnik 1* into orbit. The Americans countered four months later when they hurled *Explorer 1* into space in January 1958. Yuri Gagarin's history-making flight was still a few years off, and although neither country had launched a man into space, that didn't stop futurists from imagining the bodily rigors of interstellar travel.

Kline, who had been awarded a prestigious Lasker Award for medical research, ran the Dynamic Simulation Laboratory at Rockland State Hospital in New York, where he'd recently hired the young polymath Manfred Clynes as his chief research scientist. Clynes had grown up in Australia and attended Juilliard School of Music, where he'd become a concert-level pianist. He studied the psychology of music at Princeton University, where he befriended Albert Einstein. By 1960, however, Clynes was ensconced in the lab at Rockland, which *The New York Times* then described as "resembling the back room of a radio-television repair shop."

It was in that shop that Clynes developed one of his most lasting inventions: the computer of average transients, or CAT, a noise-canceling device for electrical brain research. By quieting some of the brain's background noise, Clynes's device enabled researchers to isolate the neural reaction to a specific stimulus, say the color blue or the sound of a train whistle. After repeating the train whistle a few times, scientists could average the neural response to determine how the brain represented, or experienced, specific colors or sounds. Clynes's machine established a direct and articulate communication system between the brain and the outside world.

The early device measured sensory input (a train whistle, the color blue) and correlated it with a measured neural response—the neural reaction evoked by a specific color or sound. The beauty of the CAT system was that it bypassed a person's conscious experience. Instead of asking a person about his experience of the color blue, the CAT system enabled researchers to observe directly how the brain responded to stimuli.

Of course, the meaning and mechanics of that neural response remained a mystery. Nevertheless, Clynes's machine gave researchers a powerful functional tool. With the CAT system, they no longer needed to rely on the muddying vagaries of subjective experience. They could observe the brain as it represented the outside world to itself.

But the CAT system did something less tangible, too. Clynes unveiled his machine in 1960—the same year he and Cline published "Cyborgs and Space"—and the machine served as an early example of the sort of symbiotic man-machine relationship their paper articulated.

The first moon landing was nearly a decade away, and exploration of deep space remained more in the realm of science fiction than actionable science. But the scientists couldn't resist the draw of interstellar travel, speculating what it would take for space travel "involving flights not of days, months or years, but possibly of several thousand years," which they warned would eventually be "hard realities."

The researchers argued that astronauts would have little chance of surviving deep space if spaceships simply mimicked the earth's environment. "We place ourselves in the same position as a fish taking a small quantity of water along with him to live on land. The bubble all too easily bursts." They proposed instead that astronauts be biologically adapted to survive an extraterrestrial environment. They called this adapted human a "cyborg," which they defined as an organism that "incorporates exogenous components extending the self-regulatory control function of the organism in order to adapt it to new environments."

In essence, the scientists were proposing a form of elective, nongenetic adaptation. They argued that space travel was not only a technological challenge but also a spiritual challenge, one that "invites man to take an active part" in his own biological evolution. "In the past evolution brought about the altering of bodily functions to suit different environments," they wrote in the September 1960 issue of *Astronautics*. "Starting as of now, it will be possible to achieve this to some degree *without alteration of heredity* by suitable bio-chemical, physiological, and electronic modifications of man's existing modus vivendi."

Their original vision of the cyborg was decidedly tame compared with the replicants Harrison Ford hunted down in *Blade Runner* or Arnold Schwarzenegger's killing machine in *The Terminator.* Rather, the term's creators believed that instead of giving an organism superpowers, a cyborg's technologies should be biologically integrated to maintain the astronaut's basic functioning—helping him breathe, process fluids, and maintain muscle tone in the oxygen-deprived and weightless environment of space.

Instead of creating a new breed of superhuman astronauts who consciously controlled their cyborg technologies, they envisioned a union between man and machine that, like the CAT system, bypassed the astronaut's conscious mind. Technologies would interact directly with the body, regulating, transforming, and performing specific bodily functions. Some of these technologies would induce hypothermia and slow the metabolism so astronauts wouldn't need as much food on board. Some would act as respiratory systems that did not rely on breathing, while others would reduce and recycle bodily waste before intravenously reintroducing it as a hydrating fluid.

"This self-regulation must function without the benefit of consciousness in order to cooperate with the body's own autonomous homeostatic controls," they wrote, undaunted by the audacity of their proposal. "If man in space, in addition to flying his vehicle, must continuously be checking on things and making adjustments merely in order to keep himself alive, he becomes a slave to the machine. The purpose of the Cyborg, as well as his own homeostatic systems, is to provide an organizational system in which such robot-like problems are taken care of automatically and unconsciously, leaving man free to explore, to create, to think, and to feel."

It was a far cry from the sleek powers of Robert Downey, Jr.'s Iron Man. Rather, the scientists had articulated a critical insight about our relationship to technology: it is through technology that we fully realize our humanity.

•

Mid-twentieth-century neuroprostheses like cochlear implants and deep-brain stimulators have already ushered in the sort of cyborg envisioned by Kline and Clynes. And while neuroprosthetists working with robot arms and cursors offered a tantalizing glimpse of what the field could achieve, it's based on what Leuthardt considered a conservative view of the body. BCIs might offer a host of new sensory experiences, stimulating the brain so that users are better able to retain information or endowing people with so-called sixth senses, enabling them to perceive portions of the visual spectrum our biological eyes cannot. Could a BCI transmit our thoughts directly to another person? How about our feelings, both emotional and sensory?

A neural augment could be just as effective in an online network as with surrounding technologies, but what happens when networks become amalgamations of man and machine—cyborg webs where machines and humans work in tandem? "I don't even like the term 'cyborgs' anymore," Leuthardt said. "The extension of our personas is no longer just the fact that we attach a machine to ourselves. It's how we network, how we can have a presence beyond our physical bodies."

With neuroprostheses, Leuthardt envisioned more than merely

manipulating the immediate environment. "So how does that change us?" he asked. "As we expand our capabilities, we potentially expand our limitations and vulnerabilities. At some point, are we going to be able to be hacked?"

The conversation between neurons and machines is a complex one and conducted in many languages: ions and hertz and amplitudes, yes, but also platinum and proteins and glia. Leuthardt knew he would need a better understanding of bioengineering if he wanted a legitimate shot at realizing his neuroprosthetic dreams. He would need access to a lab, so when he arrived as a resident at Washington University, he set up a meeting with Frank Yin, who chaired the school's biomedical engineering department. Then in his late fifties, Yin had arrived in St. Louis from Johns Hopkins University. He was a founding fellow of the American Institute of Medical and Biological Engineering. He was a fellow of the American Society of Mechanical Engineers, and he served as the president of the Biomedical Engineering Society.

He was also deeply skeptical of this young surgical resident with thick dark hair, outsized ambition, no engineering experience, and a fascination with a field that sounded as much like science fiction as it did science. Yin's graying eyebrow rose even higher when Leuthardt told him he'd studied theology in college and was now angling to spend a research year in biomechanical engineering. "Let me get this straight: you've never taken an engineering class, and you studied theology," Leuthardt recalled Yin saying. "He was exceptionally dubious about my presence." After a few tense moments, Yin told Leuthardt to talk to a newly hired researcher who could probably use a few bodies to warm up his lab.

The new scientist was Dan Moran, a turkey-hunting, tobacco-chewing, steak-eating farm boy from small-town Wisconsin who just happened to be coming off a six-year postdoc at one of the country's premier neurophysiology labs. Moran was a pure engineer—both in training and in temperament. At fifteen, he had completely reconditioned his first car, a 1966 Chevy Impala Super Sport 327. He gave the car new rings and pistons. He installed double-spring lifters in the rear so it crouched like a bull ready to charge.

But Moran's fascination with machines extended beyond mere horsepower. He'd spent his adolescence steeped in the prime-time adventures of Steve Austin, better known as television's Six Million Dollar Man, an astronaut who was "rebuilt" by government surgeons after a devastating crash. Doctors implanted Austin with a bionic eye that had an onboard zoom lens. Bionic legs meant Austin could run at sixty miles per hour, chase down cars, and kick through steel doors. His bionic arm had metal-crushing strength. It even came with a Geiger counter, because, well, as the show's narrator said during the opening credits, "Gentlemen, we can rebuild him! We have the technology! We have the capability to make the world's first bionic man . . . Better than he was before. Better! Stronger! Faster!" The show ran for only four years, but its cyborg promise left a deep impression on Moran, who collected the action figures and who to this day will occasionally sneak an online rerun or two.

Moran's childhood fantasy gained gravity a few years later, however, when his best friend injured himself sliding into home plate during a high school baseball game. It was his sophomore year, and Moran, who played catcher, was riding the bench with a broken arm. He watched from the dugout as his friend, churning hard off third, dove headfirst to beat the tag. It was a freak accident. His friend collided with the catcher's shin guards, snapping his neck. "Like that," Moran recalled, smacking his hands together. "He's been in a wheelchair ever since." His friend was paralyzed from the neck down. As Moran continued to restore his beloved Chevy, he realized he would spend the rest of his life working to transform his prime-time fantasy into a twenty-first-century reality. Moran would make his own Six Million Dollar Man. "With the Six Million Dollar Man, it was the superhuman strength, right? You can go beyond human performance, *because you're a superhero.* But when my friend broke his neck it was no longer about being a superhero. It was about getting back to normal," he said. "I realized this was a TV show—that this stuff hadn't been invented yet. But that didn't mean we couldn't invent it."

To that end, Moran spent five years as a graduate student at Arizona State University writing equations to reproduce the musculoskeletal biomechanics of bipedal motion (also known as walking), a wildly complex enterprise for anyone who has ever had to think about it. Movement for Moran was a question of torque, joint angles, force, and velocity. There was just one problem. "I didn't know anything about the brain," he said. "I'd never recorded a neuron. I'd never even seen a monkey." If Moran wanted to recreate the true nature of bodily motion, he would have to go to its source: he would have to study the brain.

It was about this time that Moran met Andrew Schwartz, who was then making a name for himself by charting how small neural populations could be correlated to arm gestures in monkeys. "I heard Andy Schwartz talk," Moran recalled, "and I was like, 'Oh, that's it. That's my guy.'" Moran did his postdoc in San Diego, where under Schwartz's tutelage he learned to open the skulls of monkeys and slide hair-thin electrodes into the motor cortex. He learned to keep the wire snug against a neuron to record its activity. He listened to their crackle and hiss, developing a series of algorithms that could transform their raw activity into a coherent set of motor commands.

When he arrived in St. Louis, Moran's tools of choice were still the slender penetrating microelectrodes he'd used with Schwartz. But Moran was a pragmatist. He hewed to the same get-'er-done ethos that had carried him through a boyhood of trapping game and customizing cars. Microelectrodes might be great for basic research—they sidled up to an individual neuron and delivered exquisite cellular recordings—but their delicate wires were brittle. They also provoked an immune response encapsulating the electrode. There was no getting around it: penetrating electrodes were simply too fickle for the sort of day-in, day-out, Six Million

Dollar Man–type brain implant Moran aimed to develop. "I can't go to a person and say, 'I'm going to give you fabulous control, but it's only going to last for a year,'" he said. "Who's going to have brain surgery like that?"

•

The joke about Moran is that while many people don't like to micromanage, Moran doesn't even like to *macromanage*. Chalk it up to a native strain of stoicism forged on the fields of Wisconsin, but Moran is what Leuthardt called a "classic 10 percenter." Say one of Moran's monkey studies winds up in the pages of *Science*. His baseline mood improves by about 10 percent. Now say a hydrogel he's working on to prevent electrode encapsulation is as effective as a jar of Vaseline. Again, the needle moves about 10 percent. Moran doesn't rile easily. He's given to wearing short-sleeved polo shirts, crisp blue jeans, and Birkenstocks or some other sensible brown shoe.

By the time Moran and Leuthardt met in 2005, the question of encapsulation seemed like one of the most urgent in all of neuroprosthetics. Moran had spent the better part of a decade implanting penetrating microelectrodes into the brains of monkeys, and with only one exception—a storied macaque known as Mojo—he had watched with dismay as his recording quality inevitably deteriorated. As Moran saw it, he had two choices: either develop a novel technique to make his microelectrodes more robust or invent a whole new way of coaxing information from the brain.

Neurosurgeons had long used a technique known as electrocorticography, or ECoG, to detect seizure focus. More recently, a handful of scientists had used the technique, placing electrode grids directly on the surface of the brain, as a research tool to record neural activity. But these were hardly brain-computer interfaces; they weren't trying to translate the information into motor commands.

So Moran had carried on, sinking his wires into the brains of rats as he built his new monkey lab in St. Louis. He was also experimenting with biomaterials, trying to develop everything from a Teflon-like substance to prevent proteins from attaching to electrodes, to anti-inflammatories that would suppress the brain's immune response. But the research wasn't going well. As in San Diego, Moran would get splendid single-unit recordings initially, only to watch them fade over time.

Moran's mood needle moved about 7 percent whenever he thought about encapsulation—something he did each time he spoke with a potential lab member. But Leuthardt was different. Like Moran, the young resident wanted to understand the language of the brain. But Leuthardt came from a clinical perspective, and while he may not have had Moran's research background, he had something else: human epilepsy patients.

"Eric was like, 'You know, I've got these human patients with grids on their brains,'" Moran recalled. "I was like, 'Really? *You've got those!*' It didn't dawn on me that every big surgical place would have these things." The idea soon emerged that they could use these temporary neural implants to decode the brain's intention. *They could use it to control machines.* Because the grid went directly on the brain, it would have greater fidelity than an EEG. By the same virtue, it didn't penetrate the brain, so it might sidestep the problem of encapsulation.

A whole new paradigm was opening before them. "Okay," Moran thought to himself, "do you want to keep hoping that people will someday solve the single-unit problem? Or do you want to get something into patients in your lifetime? The pure scientist goes with Option A. But the *engineer* goes with Option B; the *neurosurgeon* goes with Option B."

Option B it was, and they soon agreed that Leuthardt would spend the next year working along two lines of research. The first, a sort of safety study likely to bring publishable results, would use rats to investigate the brain's response to penetrating microelectrodes. But the second was pure speculation. They would use ECoG to try to produce an unprecedented brain-computer interface.

Unlike with an EEG, which attaches sensors to the scalp, their electrodes would slide below the skull. Unlike microelectrodes that pierce the brain, their platinum sensors would ride above the cortex, leaving it largely undisturbed. Better yet, ECoG would give them a distinct edge over the single-unit researchers who then dominated the field. Those researchers had spent their careers plunging electrodes into the brains of rats and monkeys. But Leuthardt and Moran would be able to get their electrodes on a substance that remains remarkably rare in neuroscience: *live human neurons*.

By piggybacking their research onto epilepsy patients, they would also avoid the thicket of federal regulations that had prevented invasive brain researchers from working in humans. The FDA had already approved ECoG grids for human use; now it was simply a matter of getting epilepsy patients (who were already planning to be implanted) to sign off on the experiments. Agreements that would take other researchers months (if not years) of back-and-forth with the FDA were simply unnecessary.

Still, no one had ever used ECoG for much more than seizure monitoring, and hospital administrators prohibited the researchers from simply running a splitter off the seizure-monitoring electrodes. That would have been a simple fix, but the hospital needed assurance that this newfangled research track wouldn't interfere with patient care. If Leuthardt and Moran were going to tap the clinical monitoring system, they'd have to do it separately, waiting until the clinical side had the information it needed.

That's when things got complicated.

Electrodes record brain waves as an analog signal. Both clinicians and researchers must digitize those signals for computer analysis. Normally, this wouldn't be a problem. Just as an MP3 can be copied repeatedly, Leuthardt and Moran could have simply borrowed the digital signals neurologists used to study seizure focus. The problem was that the clinical system digitized the signals in a way that made it impossible for Moran and Leuthardt to break them down for decoding.

They would need a work-around, a device to digitize the signals on their own. They would need, in other words, to *re-digitize* the digital signals, rendering them into a readable format.

As they tossed around various ideas for jacking into the system, they realized they could use an old brain printer—the kind neurologists once used to print brain waves on reams of paper. The printer wouldn't know where the signals were coming from. Why not feed the network's digital signals directly into the printer? It wouldn't matter where the signals came from. The printer would read them as if they were coming straight from the brain itself. Leuthardt, meanwhile, spent the next few months collecting off-the-shelf electronics as he built a second digital converter.

The idea was to feed the network's digital signals into the analog printer and then re-digitize the signals with Leuthardt's converter. It was the neural equivalent of recording a live performance on an LP, converting the album to an MP3, recording that MP3 to a cassette tape, and then re-digitizing the cassette tape as an MP3 file.

In other words, it was a total kludge. But if the system was inelegant in theory, it was downright ugly in execution. When a patient comes in for epilepsy monitoring, the analog signals come off the patient's brain, where electrodes funnel them into an amplifier and digital converter. A fiber-optic network then ferries them to the epilepsy-monitoring unit down the hall. Because the hospital's higher-ups wanted to ensure the research didn't interfere with patient care, Moran and Leuthardt had to patch into the system after the signals arrived at the monitoring unit.

As Leuthardt and Moran strategized, lining up research subjects, coordinating with Ojemann to implant the grids, and building their Rube Goldberg signal converters, they contacted Gerwin Schalk, a young research scientist at the Wadsworth Center in Albany, New York.

Some people talk with their hands. Others talk with their arms. Then there are people like Schalk who simply brutalize the air, chopping and swiping and pushing. Known as Gerv to everyone in the field, Schalk played semiprofessional football in his native Austria. Now he has the habit of standing with rounded shoulders in that slouchy, muscular way of athletes who have proven themselves on the field but now must go about the mundane tasks of daily living. Schalk's forehead is low. It's framed by a careful muss of brown hair and a deep crease that traverses the bridge of his nose. He cuts an unlikely figure for a scientist, with small blue eyes and fleshy cheeks. But if Leuthardt and Moran took circuitous routes to neuroprosthetics, Schalk hit it sideways. Growing up in the tech backwater of rural Austria in the 1980s, Schalk became a sort of proto-hacker at the age of twelve, when he began toting his Commodore VC-20 in a backpack. He spent years writing software code, building early computer models and bespoke information systems. By the time Leuthardt and Moran contacted him, however, Schalk had parlayed his computer know-how into a job as a research scientist in a prominent EEG lab at the Wadsworth Center in New York.

Like Moran and Leuthardt, Schalk was looking for a more reliable brain-machine interface. But while Moran cut his teeth on penetrating electrodes in San Diego, Schalk had learned at the feet of Jonathan Wolpaw, one of the country's foremost EEG researchers. Still, Schalk was no neuroscientist. He'd never even taken a cell biology class, absorbing what he knew instead through his work in Wolpaw's lab. What Schalk did have, however, was a deep knowledge of computer systems and programming, which made him irreplaceable in Wolpaw's lab when he designed BCI2000, a software program that converts brain signals into some form of motor output—moving a cursor on a screen or manipulating a robot arm.

The software codified the relationship between brain and machine, allowing researchers to change the parameters of both the brain waves they fed into the software and the action those brain waves produced. The software could also track external stimulation in time—having a research subject watch a cursor move to the right, for example—and correlate that visual stimulation with specific brain wave activity. It was just the sort of thing Moran and Leuthardt would need.

But Schalk had his doubts. For one thing, no one had ever used ECoG for a brain-computer interface. Schalk also came from the world of EEG, where researchers laboriously trained research subjects to achieve even modest two-dimensional control. The signals were easily corrupted—not merely by surface muscle activity, but also by something as ubiquitous as ambient electricity, which could make the signals go haywire.

ECoG would use the same techniques as EEG to pull information off the brain, but Leuthardt and Moran would have to conduct their experiments in a hospital room that positively crackled with electronics. How could their warmed-over, redigitized signals not end up corrupted? There were simply too many variables. Finally, the payoff might not be so great. "We don't even knoow if this will be aany difference in paformance!" said Leuthardt, imitating Schalk in his best Schwarzenegger accent. "We don't knoow!"

•

Unlike an EEG, where the inch-thick barrier of flesh and bone acts as a natural low-pass filter that absorbs higher neural frequencies, ECoG has no such limitation. This makes it a powerful diagnostic tool when hunting for the epileptic fountainhead. Still, anything much above 70 hertz is thought to carry very little useful information for seizure detection, so some neurologists will set their low-pass filters there, lopping off the brain's higher frequencies to focus only on the most useful information. This is not to say that ECoG delivers no information in the higher-frequency ranges—only that it's not always used clinically. But it was here, in these disregarded upper reaches, that Moran and Leuthardt hoped to find neural gold.

Coming from San Diego, Moran knew that individual neurons can fire at well above 70 hertz. Most critically, it is in these higher frequencies that neuroprosthetists have found the closest correlation between brain activity and physical actions. Could ECoG, unencumbered by low-pass filters, deliver the same high frequencies as penetrating microelectrodes?

That was the question Moran and Leuthardt sought to answer when they went to check out the epilepsy-monitoring unit. "We look at the rack, and I'm like, 'Okay, why is this low-pass filter at 70 hertz?'" recalled Moran. "The theory suggested that there should be good activity higher up." Nevertheless, neurologists assured them there was no usable information. Even Schalk, who had spent the last several years working in the lower frequencies of EEG, was skeptical.

"I came in and I said, 'Well, you guys listen to the bass, I want to listen to the treble,'" said Moran. "Not being electrical engineers, they didn't realize they could also see higher frequencies. But I came from single units. It seemed obvious that there should be some power there." Still, it remained only a theory, and the pair waited anxiously as neurologists turned off the system's filters. Sure enough, a few seconds after they removed the filters, high-frequency gamma waves came dancing along the bank of computer monitors.

Still, the real work lay ahead. They began each day unspooling a hundred-foot Ethernet cable and taping it to the floor as they snaked the cord from the epilepsy-monitoring unit to the patient's room. The cable quickly grew sticky from all the tape, but it allowed them to funnel the digital signals into the analog printer they'd set up in the patient's room. As the printer wheezed to life, it created an analog record of the brain waves before feeding them into Leuthardt's digital converter, which squatted on a nearby cart. Leuthardt's converter re-digitized the signals, sending them to a nearby PC that was running BCI2000. Schalk's software, in turn, ran the brain waves through an algorithm, producing its results in the form of a cursor on a computer screen. Meanwhile, their research subject, whose brain waves they were reading, managed to control the cursor using "only" his thoughts. "It was amazing we saw anything at all," said Leuthardt.

What they did see, however, was astounding.

Leuthardt recruited four epilepsy patients as research subjects during that first trial, spending a few hours each day doing BCI experiments. They started small. Using the same bulky electrode grids neurologists use to pinpoint seizure activity, the researchers asked their test subjects to perform three simple tasks: open and close a hand, stick out their tongue, and say the word "move."

Ojemann had placed the electrodes over the area of the brain where the temporal, parietal, and frontal lobes meet, and as the research subjects performed their tasks, the scientists concentrated on the electrodes that registered activity most closely correlated with the actions.

In the study's second phase, they asked the subjects to *imagine* performing those same actions, evoking a similar neural response. The researchers had programmed their computers to interpret increased brain wave activity as a command to move the cursor. If a research subject wanted, say, to move the cursor up, he thought of closing his right hand. By contrast, simply relaxing (rather, thinking of relaxing) the right hand would move the cursor down.

Whereas using an EEG to control a machine like this can take up to an hour of practice, the robust ECoG signals enabled their subjects to quickly gain control of the cursor. Moving randomly, a cursor would have had a 50 percent accuracy rate moving up and down. In this first ECoG run, however, the subjects accurately moved the cursor between 74 and 100 percent of the time—a success rate, the researchers boasted, that far exceeded EEG.

It was the first time ECoG had been used to gain control of a computer, but what the young researchers discovered next turned out to be even more important. During the study, they asked their subjects to use a handheld joystick to move the cursor. By tracking the brain's high-frequency signals, Leuthardt and Moran soon realized that certain frequencies were closely attuned to specific physical gestures. Move the cursor up? Neurons below a specific electrode would produce high-frequency waves. "I didn't know if it was going to be as good as single units at that point. Single units were still the gold standard," said Moran. "It turns out that high-frequency brain waves—stuff above 100 hertz—are basically an echo of single-unit activity. So we can get the same information as single units."

The experiments were more successful than any of them had dared to imagine. Moran was only just beginning his lab. He'd recently received four monkeys and funding to conduct single-unit experiments similar to what he'd done back in San Diego. But the ECoG experiment had him convinced, and as he worked with those first grants, he began using the implant chamber to record ECoG signals as well—essentially moonlighting to gather data for his next round of grants. "After that first set of grants, where we piggybacked, I never wrote another chronic single-unit grant again."

Meanwhile, Leuthardt was still in the midst of his residency. He was beginning to worry that their high-risk ECoG experiment was getting in the way of his safety experiment studying encapsulation in rodents. "This was my meat-and-potatoes project," he said. "I was really pained, embarrassed. I was like to Dan, 'Oh, I haven't got much done on the rats.' He was like, 'Screw the rats!' "

•

By then, Leuthardt was already thinking about intellectual property and inventing commercial products. He'd read a few books on the process, and he'd written a patent for a surgical retractor he'd invented. But he believed what they'd achieved with ECoG could be a game changer. "I wrote the invention disclosures and tried to get patented right off the bat," he said. Donoghue and his colleagues at Cyberkinetics had already implanted Nagle. Kennedy had his own private venture, Neural Signals, and Leuthardt was convinced that if neuroprosthetics were ever to become widely available, it would most likely be as a commercial product.

"Let's say we did this and we gave it away for free, thinking that we are doing a great thing," said Moran. "If no company has exclusive rights, no company is going to build a product because they can't protect it."

Intellectual property aside, the researchers saw another problem on the horizon: market size. "That's a fundamental practical reality," said Leuthardt. "Even though there is all this great science going on, a venture capitalist isn't going to take the risk for a small market."

And by that measure, the numbers just weren't there. The quadriplegic population of the United States is estimated at roughly 260,000, with about 12,000 new cases each year. Although there are an estimated 1.7 million amputees living in the United States, the vast majority of those are leg amputees and would not benefit from a neurally controlled prosthetic limb. "It's not a growth market," said Leuthardt. "You have to reconcile your noble intentions with a pragmatic economic reality of what the market will and will not support. How many new amputees are coming on line each year? Maybe eleven thousand? Twelve thousand? That's simply not enough for a venture capitalist. Stroke, on the other hand, is massive."

Each year, more than 795,000 people in the United States suffer a stroke. Roughly 130,000 people die annually, but for the population of roughly 7 million survivors, the results can be devastating. In the simplest terms, a stroke occurs when part of the brain dies after being cut off from its blood supply. Depending on where the cell death occurs, stroke victims can lose their ability to speak or understand spoken language. They can lose their capacity to recognize faces. They can even lose their sense of where their body is in space, leaving the patient mysteriously disembodied. But the most prevalent complication by far is weakness or partial paralysis on one side of the body. The brain is wired for contralateral control, meaning that the left side of the brain controls the right side of the body. So a stroke that occurs in the left hemisphere can result in paralysis or weakness on the right side of the body.

The country's large stroke population would present investors with just the sort of return needed to develop a commercial device. It wasn't immediately clear how Leuthardt and Moran might develop a neuroprosthetic for stroke. Nevertheless, they were convinced a solution would reveal itself, and Leuthardt moved to protect the shiny new idea. "We did that very early," he said. "You make sure you have the science down, but first you make sure you have intellectual property that is protectable."

As it turned out, just such a solution revealed itself a few years later when they discovered that neural features associated with movement were not confined to the contralateral side of the brain. Using ECoG, they were able to decode movement features on the same, or *ipsilateral*, side as well, which in a stroke victim would be the undamaged side of the brain. Would it be possible, they wondered, to create an ECoG neuroprosthetic that tapped the ipsilateral side of the brain to control movement?

Their solution is the so-called IpsiHand, which they are developing through their start-up, Neurolutions. With an EEG headset linked to a motorized orthotic, the researchers-cum-entrepreneurs hope recent stroke victims will use the device to learn to open and close the orthotic using signals from the ipsilateral side of their brains. The idea is that by thinking about moving the damaged hand while wearing the neuroprosthetic, a patient will eventually rewire his brain, training his ipsilateral side to open and close the biological hand.

An implant would give him better control, but that would entail a lengthy and expensive FDA approval process that a small start-up like Neurolutions couldn't afford. "The IpsiHand is really a beachhead to show that BCI is clinically relevant," said Leuthardt. "We can do a better job with an implant, but EEG is a low-risk entry."

What they really need, though, are good results and investor dollars, and they hope the EEG-based IpsiHand will convince venture capitalists that it's a viable product. "There's a real believability gap. It almost seems like science fiction," he said. "But as people become comfortable with it as a clinical application, it really opens the door for believability for doing an implant."

The company is now raising investor capital and conducting a small clinical trial where participants use the device to achieve one degree of freedom—opening and closing the damaged hand. Neurolutions still has a long way to go, but if the IpsiHand is successful, Leuthardt, Moran, and their cohort stand to benefit greatly: their early idea of using ECoG for a BCI is now known as U.S. Patent 7120486 B2, which names them as the sole inventors of any brain-computer interface that employs an electrode grid implanted beneath the scalp to collect ECoG signals from the wearer.

"Any BCI that uses ECoG—that's us," Leuthardt said one evening while seated at his kitchen counter and scrolling through a list of his intellectual property. "Between the surface of the brain and the scalp is what we own. Basically, if you try to decode intention using ECoG BCI, that's covered by this patent. It's a very, very potent claim."

6. THE BACKUP PLAN

Andrew Schwartz knew that if he wanted to stay relevant, he needed to sink his penetrating electrodes into human cortex. DARPA could provide that opportunity, but the agency had opted to go with the Applied Physics Laboratory at Johns Hopkins. "They have tons and tons of military contracts. So they're used to dealing with these guys," he said. "They have a comfort, and they could do all these 3-D Gantt charts, which DARPA seemed to like."

When DARPA announces a project like Revolutionizing Prosthetics, it also releases a list of potential "performers," research laboratories the agency is willing to fund as part of the project. Any researcher or lab that competes to administer a project can choose from that list, building a team across institutions. For Schwartz, that meant working with a project manager and a select group of robotics experts to build an arm before linking it to the brain. "There are less than six people in the world that really know how to build a robotic arm, and they all came from MIT," said Schwartz. "All these other yahoos basically said, 'Oh, we can build a robot arm.' You know, 'We know what we're doing.'" He added that both the Hopkins team and Dean Kamen's team talked to him about joining their DARPA proposals. "It's like, I'm sitting there: So you're going to be my boss?" he said. "Needless to say, I didn't get on any of the teams."

Schwartz was effectively locked out. The Pentagon had shut the door to humans for him, but DARPA funders were far from cutting him off. They wanted him to keep working with monkeys and awarded him $2 million for a study that not only would catapult Schwartz's research onto *60 Minutes* and into the pages of *The New York Times* but would eventually give him a shot at the human motor cortex. "They had people doing the same kind of thing that I was doing—a lot more people with a lot more money—and they didn't get anywhere," he said. "They kind of kept me as a backup plan."

Other researchers were circling around the problem of how to link the brain to a multi-jointed prosthetic limb, but few researchers had successfully closed the loop with a robot arm. Earlier closed-loop work had taken place either in the virtual environment of a computer screen or at a safe distance, as with Matthew Nagle, who performed a simple pinching action with a prosthetic hand.

Mental control of a cursor would be a boon to quadriplegics, but DARPA wanted brain-controlled prosthetic limbs—limbs you could use to brush your teeth or comb your hair. The race was on, and Schwartz devoted his research funds to a suite of experiments aimed directly at DARPA's goal: elegant neural control of a dexterous multi-jointed limb. "It turned out to be great," he said. "I didn't have to report to APL or anybody. I just did my own work."

With electrodes in hand, Schwartz and his colleagues began work with two monkeys and a pair of robot arms outfitted with a pincerlike claw instead of a hand. Training research monkeys falls somewhere between art and science. Since you can't tell a monkey what to do, researchers must devise ingenious ways of familiarizing the animals with the physical essence of a task. It's a delicate procedure, and Schwartz began by training his monkeys to control the arms using a joystick.

Pressing the joystick forward, the animals learned they could extend the limb to various fixed points in space, grab a marshmallow or a sliced grape from a skewer, and pull back on the joystick to bring it to their mouth. As the monkey brought the marshmallow back, researchers impaled the next food reward, fixing it in one of four positions for the animal to grab. Once the monkeys were familiar with the task, researchers removed the joystick, immobilizing the animals' arms by placing them in tubes attached to the task chairs. Meanwhile, they recorded neural activity while placing the arm under "automatic control," giving researchers command over the arm as it grabbed food and brought it to the monkey's mouth.

One of the great discoveries of the late twentieth century happened in the lab of the Italian neurophysiologist Giacomo Rizzolatti. The scientist had implanted electrodes in the pre-motor cortices of monkeys, hoping to listen in on neurons he believed were associated with hand and mouth movements. The researchers recorded from individual neurons as a monkey reached for a peanut, tracing the cell's firing pattern before, during, and after the movement. By that measure, Rizzolatti's experiment did not differ tremendously from the neural recordings his fellow researchers were making in other labs.

What set Rizzolatti's work apart, however, occurred by accident. During a break between tasks, the monkey sat idly in its chair as researchers milled about the room. The monkey wasn't moving at all, but when one of the researchers snatched a spare peanut and popped it in his mouth, the neuron they had been recording erupted as if the monkey had grabbed the peanut itself. It was a shocking discovery: the brain, or at least a specific class of cells, seemed not to distinguish between an action performed and an action observed. Here was a class of neurons, later dubbed "mirror neurons," that was involved in motor planning but that was also interested in the physical actions of others.

Much has been written about mirror neurons, and brain researchers such as Marco Iacoboni at the University of California,

Los Angeles, have proposed that the mirror-neuron system plays a critical role in recognizing the needs of others. We flinch when we see someone injured on the street. We thrill at most any sport, and we feel deep sympathy for the fictional trials of characters in film and theater. Why? Because at some level our brain physically re-creates the experience as though it were our own. Mirror neurons, these researchers believe, not only are the fundamental mechanism by which we feel empathy but also play a role in so-called theory of mind, enabling us to recognize that other people have ideas and desires that are distinct from our own.

At a more practical level, the brain's penchant to re-create observed actions helps researchers such as Schwartz to prepare a monkey's brain for BCIs. As the monkey watched the arm grab a piece of food, the animal's motor neurons began firing as though it were grabbing the fruit with its own biological arm. Meanwhile, Schwartz and his colleagues used the information to build their "decoder," the computer algorithm that associates specific neural firing patterns with particular movements. As the researchers continued moving the arm, the algorithmic association between firing patterns and the robot arm movements grew stronger.

Eventually, they began to dial down their control of the arm, blending automatic control with signals from the animal's motor cortex. The monkey had partial command of the arm, but scientists could still correct its movements if the arm began to go wildly off course. They had effectively given the monkey training wheels, encouraging it to move the arm in the desired back-and-forth direction but constricting the arm's movement from side to side. It was a synergy between animal and algorithm: the computer was learning to better interpret the monkey's neural code; the monkey's motor neurons were learning to better control the arm.

•

Earlier center-out experiments gave animals a fixed beginning and end point to their tasks. The monkey knew to begin the task when the cursor appeared in the center of the monitor. Once the animal successfully moved the cursor to a target, it would receive its reward, and the task would reset. The animals needed only to control the cursor for the time it took to move it from the starting position to an end target, sidestepping the added complexity of controlling the cursor for long periods of time as it traveled home. Schwartz, by contrast, gave his monkeys continuous control of the arm. Unlike earlier studies, the animal didn't return to a "home" position after grabbing and eating a food reward. Rather, the monkey had to maintain control of the limb as it reached to a new location for the next reward.

It was a complicated task, but again Donoghue had beaten Schwartz to the punch. Using the BrainGate system, Matt Nagle had not only managed a pinching action with a prosthetic hand but also gained continuous control over a computer cursor, using it to navigate a desktop environment, opening e-mail and playing music. "He could hardly move that cursor on the screen. It was terrible," Schwartz said of Nagle's shaky performance. "They point to my stuff, and they say, 'Well, you can get it to work in monkeys, but it won't work in a human.' " Donoghue's performance might have been unconvincing to Schwartz, but there was no denying that he'd granted a human research subject continuous control of a computer system.

What Schwartz's research lacked in novelty, however, it made up for in sophistication. Using his brain-computer interface, Schwartz's monkeys gained four degrees of freedom to perform an essential function: self-feeding.

Perhaps even more important were the behaviors that began to emerge. Rather than using a neural cursor, which exists only in the virtual realm, Schwartz's monkeys were directly interacting with a physical limb. In a video of the research, one monkey sits in a task chair, the view of its protective resin cap obscured by aluminum scaffolding. The monkey's arms are immobilized in plastic tubes. The black ribbon wire that emerges from its skull peeks from behind the scaffolding whenever the animal strains it neck. Meanwhile, the industrial-looking robot arm, a device of shiny steel tubing and a pincer wrapped in white fabric, methodically reaches to grab marshmallows. The monkey flexes its right arm as the robot reaches for a marshmallow. Bringing the food to its mouth, the animal strains toward the marshmallow, successfully completing the task seven times in a row.

Researchers sampled the monkeys' brain signals every 30 milliseconds, giving them about thirty-three snapshots per second of the monkey's neural activity. It took the algorithm and robot arm about 150 milliseconds to transform each neural impulse into an arm command, which is about the lag time of a biologically intact nervous system. In essence, Schwartz and his colleagues had given the animals instantaneous and uninterrupted control of the arm.

Nowhere was this more apparent than in a second video when a researcher moves a piece of food away as the monkey is in midreach. The animal quickly changes course, grabs the marshmallow, and takes a bite. In a third video, the monkey only partially takes a marshmallow into its mouth. With its biological arms restrained, it holds the marshmallow in its lips, using the pincer to push the treat all the way into its mouth. It wasn't making rote, stereotypical movements. It was freestyling.

Of course, a biologically intact person continually adjusts her movements to better adapt to the shifting environment. By that measure, the monkey's movements are banal. But that's really the point—a smooth, naturalistic movement that approaches its biological analogue. The monkey was coming to "embody" the robot arm, its neurons spontaneously adjusting to better achieve the task at hand.

At that moment, Schwartz's monkeys were doing something not so different from what a baby does when she learns to walk or from when a child learns to throw a ball. But while biological coordination improves as pathways between neurons are streamlined and strengthened, the neurons here were changing their behavior to better control the nonbiological arm. They were adapting to the interface.

Meanwhile, Schwartz had used lessons from his early days with Georgopoulos to craft a relatively simple algorithm that could recalibrate to accommodate the dynamic neural interface. "It's important to realize that we have a model. Our model is far from perfect in describing what really goes on in the brain," he said. "But what we've been incredibly successful at is extracting movement and intention by using this very simple algorithm."

•

The resulting paper, published in 2008, was a watershed moment for the lab. CBS's *60 Minutes* came calling. The study landed on the front page of *The New York Times* and was subsequently picked up by countless other news organizations. No one had ever shown such elegant neural control. Schwartz had clearly knocked it out of the park, and his lab was inundated with interview requests. It was a gratifying moment for Schwartz, but not a comfortable one for a guy who's more interested in the science than in the demo. "I hated it. I could never express what I wanted to express. All I could say is 'self-feeding. Yeah, they can grasp pieces of food and bring it to their mouth,'" he said. "You end up telling the same damn thing over and over." Still, Schwartz was undeniably proud of the work. He'd shown proof of principle: not only could a monkey gain elegant and continuous control over a robot arm, but it could also use it as a worthy surrogate of its biological counterpart to perform an essential task.

The whiz-bang factor of BCI kept the public interested—and, importantly, the cash flowing—but it was the underlying science that most excited Schwartz. Brain-computer interfaces were pointing to some foundational principles of brain function. They were telling him things about how the brain learns, its relationship to objects, even thought itself. "I always laugh about psychologists and cognitive neuroscientists who say they're going to study cognition or thinking. I say, 'Can you define that for me? If I were going to poke one of my electrodes in the brain and find a thought, how would I know if I found it?' " he said. "They can't define it! They can't even define the necessary parameters of thought, so how am I supposed to find it?"

What Schwartz had developed, by contrast, was a closed input-output system he could use to test the accuracy of his model. "We can prove how well it works because we can look at the movement, or the performance. You can't do that if you say, 'Oh, thought takes some electricity and some chemicals.' Where's your model?" he said. "But I can say, well, based on my model—my hypothesis—my subject can do this."

The self-feeding task didn't give Schwartz a way to explore the more fundamental questions of how the brain generates neural code or why a motor neuron changes its activity pattern. What it did give him, however, was a way to observe the brain as it shifted those patterns of activity.

Simply stated, an individual motor neuron will fire more rapidly when initiating movement in a "preferred" direction. The farther the intended movement is away from an individual neuron's preferred direction, the more slowly that neuron will fire. It was by combining the firing patterns of a population of individual neurons—a population vector—that the Georgopoulos lab first accurately anticipated movement.

The field has been working with that model since the mid-1980s. But one of the features Schwartz and others have highlighted since is the tendency of neurons to shift their preferred firing direction to better accommodate external modalities such as robot arms.

A key tenet of neuroplasticity is that the brain reorganizes itself by creating fresh synaptic connections between neurons. By forming connections between cells, neurons "wire" together. Their activity becomes more closely associated. When one wired neuron fires, a small pulse of electricity courses through the cell body to where it forms synaptic connections with other cells. The pulse of electricity causes the cell to release chemicals (known as neurotransmitters) into the gap separating one cell from another—a synaptic connection.

Each of the brain's estimated 100 billion neurons is synaptically connected to an estimated 10,000 other neurons. At any given moment, an individual neuron may be receiving inputs (in the form of neurotransmitters) from thousands of neighboring cells, each coaxing it to produce or withhold an action potential. Once the receiving neuron reaches an informational threshold, it will produce an action potential of its own, releasing still more electrochemical signals to nearby neurons (each of which is receiving inputs from thousands of other cells).

No one knows for certain what the tipping point is for a neuron to fire. Is it an accretion of inputs from thousands of nearby neurons? Are there certain neurons whose activities are so intimately bound that when one fires, the other does as well? Do some neurons have more influence than others? Are they all equal? No one really knows.

Nevertheless, new synaptic connections are critical: it is a physical alteration of the brain's physiology to produce new behaviors. Said differently, it is the physical process of learning. We learn new behaviors or skills by altering our brain's activity and physical landscape, and these changing synaptic connections are the fundamental building blocks of that process.

And yet no one understands the underlying mechanism of this process. "Everybody talks about synapses changing their efficacy—that they're plastic and that their synaptic efficacy changes, but that's not a model. That's not showing you that this happens, and then this happens, and I get this result," said Schwartz. "But with BCI, we can do that. We can actually make a subject learn."

Schwartz can induce changes in how the brain behaves. "I can explicitly force you to change the way your neurons fire," he said. By altering the output algorithm that controls the arm, Schwartz can make a neuron whose activity is normally associated with, say, moving the arm up and to the right initiate a movement in another direction. Faced with such a contradictory output, neurons in the motor cortex will actually change their directional tuning to accomodate the new paradigm. "That is much closer to the way learning really takes place in the brain than trying to understand how some neurotransmitter changes a little bit or how a protein changes," he said. "I can't tell you *how* a neuron changes its tuning function, but I can tell you certain *ways* that it changes its tuning function."

•

DARPA was more interested in results than in Schwartz's basic findings, but the motor control he exhibited meant he was on the inside track to humans. Not long after he'd published the 2008 paper, Kaigham Gabriel, then the acting head of DARPA, called asking to visit Schwartz's lab. APL had already produced two prototype arms, but direct cortical command was a long way off. "They had thirty or forty labs, but they didn't know what they were doing," Schwartz recalled. "I was out of that whole Revolutionizing Prosthetics thing, but with the results I was getting, you know, my little lab by myself—we got all these results that nobody else was getting."

When Gabriel touched down in Pittsburgh, Schwartz gave him the grand tour. He showcased the monkey lab and the self-feeding experiments. "They spent all this money, and they really didn't have anything," said Schwartz. "But for me? This is my career. This is my life."

Schwartz squired the DARPA head across town, where his department chair, Michael Boninger, gave him a tour of his wheelchair lab for spinal cord injury patients. "He knew we had this capability," Schwartz said. "We could do this stuff." Gabriel was impressed, and by the end of the visit he floated the idea that Pittsburgh might join the Revolutionizing Prosthetics program. "I'm going, *Whoa*!"

By then, Donoghue had already been working in humans for

years. Schwartz had made tremendous progress in monkeys, but he was getting impatient.

He needed to get into humans. Fast.

For the past few years, his former postdoc Dan Moran had been working in St. Louis, where along with Eric Leuthardt he had been pioneering an ECoG-based brain-computer interface. Moran had been proselytizing about ECoG's potential ever since he and Leuthardt first hatched the idea. But Schwartz was skeptical. He was a purist. He loved the clarity of a single neuron, and ECoG just seemed muddy. "You don't have enough information coming out of that signal," said Schwartz. "The resolution is just too low to get the details you want." But Moran, Leuthardt, and their collaborators had forged ahead, publishing a string of promising results.

Flush with this success, Moran was now visiting his mentor in Pittsburgh. "When I told Andy that Eric and I were going to start doing ECoG back in 2003, he'd said, what the hell are you doing wasting your time with that crap for?" said Moran. "But it was a great way for him to get the infrastructure for human experiments. I think he realized ECoG is safe, so he can get into human patients. It's more pragmatic."

Pragmatic? More like quick and dirty. Unlike the penetrating electrodes Donoghue had pushed through an onerous FDA approval process, ECoG grids were already in wide clinical use. The FDA didn't need to be involved. All Schwartz and his department chair, Michael Boninger, needed to get up and running on a human ECoG trial was approval from the school's Institutional Review Board.

"I'm thinking I've got to do human stuff," said Schwartz. "We're talking, and Mike said, why can't we start with epilepsy just like you guys are doing in St. Louis? I wasn't super excited about it, to tell you the truth. I looked at it as infrastructure." ECoG would let them start working in humans immediately, but what he really wanted—needed—was to sink his penetrating electrodes into the human motor cortex. "Why can't we do that?"

Boninger recalled asking over a beer. "We were talking, and Andy said, 'Man, if I could get these arrays in someone, I could get them to play the piano.' "

•

The summer of 2004 was a sort of golden moment for Tim Hemmes. It had been a few years since he tabled his high school dream of playing professional hockey. Now twenty-three, he planned to marry the mother of his eighteen-month-old daughter. His fledgling auto-detailing business had survived its first winter. Business was starting to pick up, and he'd worked out agreements with several used-car lots around town to refurbish vehicles they bought at auction. He was making good money for the first time in his life. The business was booked solid for a month. He'd even hired a few new employees.

His girlfriend was studying to become a nurse, and they'd started talking about wedding dates. They were on the market for a house, and he'd just bought his first motorcycle, a blue and white Suzuki GSX-R750, a beast of a bike that topped out at 171 miles per hour. They'd even adopted a new pit bull puppy that he'd named Neo after Keanu Reeves's character in *The Matrix*.

Hemmes had always hoped to get out of the small towns around Butler, Pennsylvania, but the place had a way of sucking you in. He hadn't had the grades to go to a good college. And now that his NHL dreams hadn't panned out, Butler didn't seem so bad. He was starting to think about his five-year plan. He was even starting to think about his ten-year plan.

It wasn't what he'd imagined, but his life was coming together, and on a sunny day like July 11, 2004, his future seemed clear, bright, and untroubled. It was one of those Sundays when Hemmes had several plans in the air. He had talked about spending the day in Pittsburgh, grabbing a bite to eat and touring around town. There was also a concert that night. But his girlfriend's sister had wanted them to babysit her son, so they instead loaded up the car and headed to the park. The temperature was on the gentle side of summer, a slight wind and not a cloud in the sky, and when they arrived at the park, Hemmes folded his six-foot-two-inch frame onto a bench to watch his daughter run around with their new puppy. "That was my day," he said, "everything I could've asked for."

They fixed something to eat when they arrived home, but it was a gorgeous night, and Hemmes was restless to take his bike for a ride. He loved its lines and its power. Though he'd only had the motorcycle for a few months, he'd already put more than two thousand miles on it. His friends had cautioned him against getting such a big bike his first time around, but that had never made sense to him. He'd wanted a big bike. He already knew how to ride. He also knew himself well enough to know that if he bought a smaller bike, he'd just end up buying a bigger one a few months down the road. Why waste the money? The bike's slogan was "Own the Racetrack," but Hemmes wasn't looking for speed that night. He just wanted the summer wind in his face as he cruised the two-lane blacktops outside Butler.

Dressed only in a pair of jeans and a long-sleeved T-shirt, he was caught short when he pulled out of the garage. "Wear your helmet," a voice said, prompting him to turn the bike off and head back to the garage.

It was around 10:40. Hemmes had been traveling these roads his entire life. He wasn't going more than twenty miles per hour, but when a deer jumped in front of him, he swerved hard to the right, hitting a patch of gravel. He touched his brakes, trying to pull out. But the pebbles were like ice. Hemmes was skating on two wheels, gripping hard on the handlebars as he careered toward a mailbox by the guardrail.

The bike was still in first gear, but Hemmes hit the mailbox with such force that he pulled its post from the ground. The impact snapped his right scapula in two. As he fell back, he twisted the throttle, unleashing the engine's eighty-three newton meters of torque. The bike reared on its back wheel. It jammed its tailpiece into the ground with such force that it kicked back like a donkey, popping upright and launching him over the handlebars.

Hemmes awoke in a hospital three days later. He had a tube up each nostril and one down his throat. His feet were elevated. His head was low, and the room was sweltering from all the heart monitors and ventilators whose beeps and chimes and dings filled the chamber. "I'm thinking, why am I in this room? There was no one there to explain," he recalled. Then he tried to scratch his nose. "I was lifting my arm, I could feel it in my mind moving," he said, "but it wouldn't move. I had no idea why."

The accident had shattered Hemmes's fourth cervical vertebra, severing his spinal cord and leaving him paralyzed below the neck. "There were no bones, no anything supporting my head, just the meat and tissue," he said. "I was literally a bobblehead." Doctors tried to stabilize the spine, but his C4 vertebra had turned to dust. They ended up plucking bone fragments from his neck, implanting a bone from a cadaver, and using rods to fuse his cervical spine from the third to the sixth vertebra.

Through it all, Hemmes refused to believe he was paralyzed. He told himself he was on vacation, just lying in bed watching television. "That's what I had set in my mind. You know, I work very hard, and this is going to heal itself," he said. "It may take six months, it may take a year, but I was on vacation." Still, the ventilator tube made it hard for him to talk, so his family wrote out an alphabet on a piece of paper. They would run a finger over the letters. He would nod yes to spell out words.

"This sucks," he spelled out one day when his girlfriend came to visit.

She asked him what he meant.

"I want to write something," he spelled.

He couldn't, she told him. He couldn't hold a pen.

He nodded yes. Yes, he could write.

"No, you can't, Tim," he remembered her saying. "You're paralyzed."

She placed a piece of paper on the bed beside him and placed a pencil between his fingers. Hemmes tried to grasp it, but the pencil fell to the floor. She picked it up and again placed it in his hand. Once again, it dropped to the floor.

"Then it finally hit me," he said.

•

Hemmes's muscles atrophied in the coming months. He wasn't eating. He dropped to a mere 150 pounds, as he grew increasingly despondent. He might have remained that way if he hadn't seen a TV show about an experimental stem cell treatment in Portugal. "She was walking," Hemmes recalled of the patient. "My mother hadn't seen the segment, but when she came back in the room, I kept saying, 'I need to go to Portugal. I need to go to Portugal.' "

Hemmes never did go. The clinic only harvested a small number of stem cells, and his injuries were simply too severe. He kept hunting, though, and in November 2006 he underwent a similar treatment in Mexico, where doctors removed a band of scar tissue around his injury before implanting some two million embryonic stem cells, an experimental treatment not approved by U.S. regulators.

Hemmes emerged hopeful from the procedure. The surgeon told him the operation had been a success. Hemmes recalled that the doctor said he'd found 40 percent of the spinal cord still intact. Scar tissue had been choking it off like a tourniquet, and once he removed the constricting tissue, the cord "immediately turned back to the right color and started to pulsate like normal." Hemmes spent his twenty-sixth birthday at the Mexican clinic, where doctors treated him like a friend, giving him birthday gifts, buying him beer, pizza, and cigarettes, while nurses stayed up late with him watching Italian films and brought him a slice of double-chocolate cake.

He was buoyed a few days after the surgery, when one of the doctors injected him with antibiotics. Hemmes couldn't believe it. He hadn't felt anything below his neck in more than two years, but he immediately felt a sharp, burning sensation. "No way am I feeling this injection," he thought. "But when I looked down, right where it was burning, he was giving me my shot." The doctor then removed the needle, Hemmes recalled, and said, "You will walk again."

Hemmes never did walk again. But that was never really the point. "It was a fantasy," he said. "I'm not looking to walk out of these places. I'm looking for a little bit of recovery. What a lot of people don't understand with spinal cord injury is that if I could just move my little finger, that would change my world." Hemmes had put in the hard work. He'd gone to physical therapy. He'd tried alternative treatments and even traveled abroad for experimental procedures. It was time to accept his injury. It was time to concentrate on building a life and raising his daughter.

Nevertheless, he continued to read up on experimental treatments and research. He signed up as a potential research subject with the University of Pittsburgh Medical Center. He was also talking to researchers in Cleveland who for the past three decades have been developing a technology known as functional electrical stimulation, or FES, which uses electrodes implanted in the body's peripheral nervous system to reanimate paralyzed limbs.

By then, Schwartz and his colleagues were looking for human research subjects for an ECoG study that would be headed by a young researcher named Wei Wang. "They were telling me about how I'd be able to move a robotic arm and a cursor on a computer screen. They told me it would be no direct benefit to me but that it was going to be the foundation for something bigger," he said. "Knowing my luck, I'd say yeah to the Pittsburgh study, and that's when Cleveland would call, and there was my opportunity for FES out the window."

Hemmes said no to Pittsburgh. But a few days later, the researchers called again. They'd spoken with Cleveland. They told him there might be an opportunity, years down the road, to combine an implanted wireless neuroprosthetic with a wireless FES system. He wouldn't be just controlling robotic limbs, he would be able to use his own. "Now you got my ears open," he said. "I would be their guinea pig for the next five, ten, or fifteen years. I'll do whatever surgery needs to be done, but when that time comes and we do the complete wireless system, I want to keep it. You can't take it out of me."

•

In the late summer of 2011, Hemmes became the first human subject to receive an ECoG implant outside an epilepsy-monitoring setting. Hemmes would have the implant for no more than twenty-eight days. Time was short, and a mere two days after the surgery Wang and his fellow researchers arrived at Hemmes's home to connect his brain to a computer.

Surgeons had slid the implant below the dura mater and then routed the connecting wires down his neck to his chest, where the cables exited his body and connected to an amplifier. His head was pounding. Stitches still bound the incision, but on that first day of testing, Wang and Hemmes worked to gain two-dimensional control over the cursor.

Unlike Leuthardt and Moran, who asked their human research subjects to imagine distinct physical gestures like making a fist to move the cursor to the right, Wang first asked Hemmes to imagine moving his arm naturalistically to control the cursor. "They kind of left it up to me," he said. "If I wanted to go up, I would think lift my arm up. If I wanted to go down, I would think swing it down—left, right, all the same thing." The grid's electrodes were each listening in on different groups of neurons, meaning, theoretically, that the implant could provide nearly thirty independent output channels.

That would provide ample information to gain control of a cursor. Still, Wang was nervous. He realistically had only a few weeks to work with Hemmes, and Wang, an assistant professor heading a closely watched study, needed results. "Very stressful" was how he recalled the monthlong experiment. "It's almost like a space mission: you prepare all of these things up to a point, but then you launch the thing, and you only have a certain amount of days to complete the mission." Wang wanted to demonstrate sound three-dimensional control in a human. By then, Leuthardt and his colleagues had already shown two-dimensional control, but time was running short, and Wang discarded the naturalistic yet more complicated approach of asking Hemmes to imagine moving his arm to the left or right.

They asked him instead to imagine unrelated actions to move the cursor. Hemmes would think of moving his thumb to send the cursor to the left. He'd imagine flexing his elbow if he wanted to move it to the right. Imagining moving both the thumb and the elbow sent the cursor skyward, and so on. Not only did the system enable Hemmes to move the cursor in three dimensions, but by modulating the speed of the imagined movement, Hemmes could simultaneously adjust the cursor's velocity, eventually reaching an 80 percent success rate.

It was no mean feat. Hemmes had to remember and coordinate six unrelated actions to control the cursor in just three dimensions. "I'm trying to get up in the top corner," he said. "I'm trying to move my thumb just a little bit, because if I do too much, it's going to fly backwards while I'm trying to lift my arm while trying to make my fist. There was a lot going on."

It was a cumbersome system, and Hemmes relied on the skills he'd developed as a hockey goalie, keeping his eye on the puck while paying attention to other players on the ice. "You're always making those calculations," he said. "It's the same with this; there are so many different things going on at once."

Even so, this sort of system, known as a "classifier," seemed inherently limited. It's hard to imagine that a person could gain much more than basic three-dimensional control using a system of imagined gestures to enact unrelated movements. "The current study was limited by . . . the relatively arbitrary association between attempted movement and desired cursor movement direction," Wang noted in *PLoS ONE*, the online scientific journal where he eventually published his findings. "It is worth investigating BCI control schemes based on natural neural representation of intended movement in ECoG signals."

Worth it? For Schwartz, it was an absolute necessity. If neuroprostheses were ever to become truly useful, they'd have to use an intuitive form of control. A classifier was a quick way to gain simple control, but it would ultimately become a limitation that prevented more complicated and spontaneous control, which soured Schwartz on the potential effectiveness and scientific relevance of ECoG. "Every time I see Wei, I say that's bullshit," he said. "That's a classifier. You don't need to know anything about how the brain works other than the fact that when the guy does what you tell him to do, you can make those neurons fire."

Schwartz argued that because a classifier relies on a limited collection of arbitrary gestures, it would never allow a person to move creatively and spontaneously. A user would instead be stuck with a handful of rote actions, and there was no guarantee he or she could extrapolate subtle movements that fell outside that limited vocabulary. "I kept telling them this is not the way to do it," Schwartz said. "It's an engineering shortcut. They're doing engineering without a foundation in science."

Nevertheless, on the final day of the study, Wang hooked Hemmes up to a robotic arm. With cameras rolling, Hemmes used the arm to reach out and give Wang a high five. "I was able to go to him," Hemmes said. "That was the first time in a little over seven years that I was able to interact with another human being in that way."

•

As impressive as Hemmes's feat was, he couldn't have achieved it without the latest version of APL's modular prosthetic limb, a technological marvel that boasted twenty-six degrees of freedom and sensors to transmit sensory information to the wearer.

Still, to fully exploit the arm's potential, researchers would have to link it to the central nervous system. That meant DARPA had to find a scientist with the research know-how to control the arm. That meant DARPA needed Schwartz.

"I said this is what we're going to do: boom, boom, boom, boom, boom," Schwartz recalled. "This is exactly what I wanted to do, and it takes a guy like me to make it work. It's like, I don't give a shit if you give me money or not; I'm going to make the damn thing work—and that's what it takes."

Once Geoffrey Ling saw Schwartz's proposal, the choice seemed clear. "He totally understood what we wanted to achieve," said Ling. "Because of who he is, and the people he has working with him, he was clearly the person."

7. FEELING THE LIGHT

"Ninety-nine percent of the field is trying to control upper limbs. These guys are obsessed because that's the only thing they know how to do," Miguel Nicolelis said while seated in his office at Duke University's leafy medical school campus. "That's the same thing we did in 2000. It happens a lot in science: there is one idea, and everyone tries to follow that idea. Our lab creates these ideas—many, all at once. Our idea is to go way beyond that."

It was the summer of 2012, and outside Nicolelis's window construction crews were noisily erecting a new glass-and-steel structure amid the hilly campus's evergreens and oaks. Inside his office, however, the neuroscientist was thinking about soccer. That year's Brazilian team was being compared to the storied team of 1970, which under the likes of Carlos Alberto and Pelé was perfect during its World Cup victory. The team was playing in the Euro cup, and Nicolelis was planning to stay up late to see the game, which he'd been previewing on his iPad.

Nicolelis first developed his passion for soccer as a child in Brazil, where he studied medicine before immigrating to the States to work with the physiologist John Chapin. In the summer of 2012, however, Nicolelis was thinking about soccer in an entirely different way. Namely, he was working with an international team of scientists to build a full-body, brain-controlled exoskeleton. The plan, as Nicolelis envisioned it, would be for a quadriplegic to don the exoskeleton during the opening ceremonies of the 2014 World Cup in Brazil. Rising from a wheelchair, he or she would use the exoskeleton to walk to center field and kick a soccer ball before the first match. "The opening kick would be a demonstration that science can almost do the impossible—make someone walk again," he said. "It would be the only soccer game in Brazilian history where nobody remembers the result of the game because they would be more fascinated by what happens at the opening."

To that end, Nicolelis had devoted several bays in his vast monkey lab to what he calls the Walk Again Project, which he believed would move the field beyond upper-limb prosthetics to enable full-body control. Nicolelis and his collaborators were working to expand the recording capacity of their electrodes, moving beyond the Utah array and instead implanting research animals with hundreds of electrodes across multiple brain regions. "We need 141 degrees of freedom for that exoskeleton to be fully operational—legs, arms, fingers, everything. You don't do that with two hundred neurons," he said. "Once you start getting ten thousand or twenty thousand neurons recorded simultaneously—this is going to change the game. Because you're not talking about seven degrees of freedom: you're talking about tens of degrees of freedom."

A short walk across campus, Nicolelis's monkey lab is a testament to the scientist's success. The one-story building is clad in iconic Duke stone—a rough-hewn slatelike material flecked with blues, tans, and rusts quarried in nearby Hillsborough, North Carolina. Whereas most labs rely on other departments to perform implantation surgeries, Nicolelis's lab boasted a dedicated operating room for the procedure. Located near the lab's entrance, the room was built for monkey-sized patients with a small stainless-steel operating table at its center. A large microscope stood to the right of the table, and various contraptions—an EKG machine, heating pads, an oscilloscope—sat on a counter running the length of the room. From the ceiling hung a pair of surgical lamps, while shelves housed an assortment of Huggies diapers and catheter tips.

Nicolelis's lab is U-shaped, with individual observation rooms just outside the interior monkey bays. Sock puppets, black lights, Clorox wipes, and an assortment of what looked like retired cables, electrode strips, and wires littered the work space, where under Nicolelis's guidance researchers have performed some of the most radical BCI experiments on the planet.

In one bay, a monkey named Cherry was running a basic center-out task, moving a virtual computer arm from a home position to a randomly selected target in another area of the screen. The twist here was that Cherry was not controlling just one arm. She was using nearly eight hundred electrodes implanted in ten separate areas to control two arms simultaneously. Each arm had only two degrees of freedom (for a total of four degrees). So far, her accuracy rate was around 80 percent.

As Cherry worked, a graduate student named Peter Ifft manned three computers studying her progress from the observation room. Two of the computers were devoted to recording Cherry's brain activity. The third funneled those recordings through an algorithmic maze—calculating as many as ten thousand neural spikes per second—and translated them into virtual arm movements. The room's five monitors showed grid after grid of individual electrode recordings, which formed undulating waves of blue, yellow, and pink representing individual neurons each electrode shaft recorded.

An audio monitor crackled with the sound of Cherry's humming brain, while above Ifft's work space a black-and-white monitor showed the animal sitting in a metal chair.

The grainy image showed Cherry with what looked like a plastic halo crowning her head. It was actually an enclosed cup that housed her implants' electronics. The implants extended her nervous system via a thick bundle of rainbow-colored ribbon wires that flowed from her cranium, up to the ceiling, and into the observation room, where they cascaded into Ifft's computers. "We're scaling up the system. It all has to do with the more complete BMI—a full-body BMI," said Ifft, a tall man dressed in a white full-body lab suit. "We're aiming for a two-limbed BMI, which is another level of complexity. More cells, more channels, more quality recordings, should enable that."

Cherry's microelectrodes were spread across various regions of her brain—not only the primary motor, pre-motor, and sensory cortices, but also higher-order brain regions associated with cognition and decision making. The idea was that by listening to these regions communicate with one another, the Nicolelis lab could re-create more refined and complex motor movements across both arms, which they would eventually integrate into a full-body BCI.

"Scaling up" to two arms wasn't so straightforward as simply implanting electrodes in both motor cortices. Rather, the Duke team had found cells in other brain regions that were inactive when Cherry used only one arm but sparked to life when she used both. The trick was creating a BCI that understood which movement Cherry wanted to make—simply grasp a jar, or grasp a jar and open its lid—and coordinate the movements. It became even more complicated at the cellular level: whereas neurons in the motor cortex were directionally tuned for one-armed movements, the cells' tuning properties shifted when the movement was integrated with a second arm. A truly functional bilateral BCI would have to understand this higher-order state, determining whether the user was trying to move one or both arms.

Eight hundred neurons delivered ample information for Cherry to control two arms, but Nicolelis was convinced they'd need many more for a full-body exoskeleton. "It's going to cross the threshold for control that can be useful for patients," he said. "That's the key issue here. If you only have twenty to fifty neurons available in your recording, forget it. You're not going to do anything meaningful."

A big part of that issue is going beyond mere voluntary movements of the arms and legs. An exoskeleton would also have to mine the brain for subtler control functions like gait and balance—aspects that are thought to be subcortical, buried beneath the neocortex in some of the brain's older structures. But here, Nicolelis was convinced he'd already found a solution. "The same pools of neurons that control the legs can produce information about the posture of the animal," he said. "So we have posture control in monkeys from the motor cortex, which is a big breakthrough."

Enabling this sort of subtler neural control was the order of the day one monkey bay down, where researchers were testing a miniature prototype of the exoskeleton. The prototype comprised only a padded pelvic girdle and a pair of leg braces outfitted with pneumatic pistons. Researchers had suspended the monkey-sized exoskeleton from an aluminum frame, which had a collar extending from its top beam to secure the animal in place. The entire apparatus was built around a treadmill, which researchers were using to simulate bipedal motion.

Macaques don't naturally walk long distances on their hind legs, but after a bit of wrangling the researchers managed to strap a large monkey named Mango into the exoskeleton. Using Velcro straps to secure his legs, they restrained Mango's upper body with the throat collar. The monkey was still being trained to walk upright, and he looked a little confused in his new rig. His tongue stuck out slightly between his teeth, as his torso and legs—naturally given to a quadrupedal crouch—were stretched vertically. His feet slightly pigeon-toed, they just barely touched the treadmill as researchers began applying pieces of reflective tape at his hips, knees, and ankles. Black lights were placed in a ring around the frame, and using a modified Microsoft Kinect, the researchers planned to track the monkey's gait.

Once Mango was situated, the researchers closed the door to the monkey bay, while a graduate student turned on the pneumatic pump. The motor sputtered to life with a staccato hiss as it fed tiny bursts of air to the exoskeleton's pistons, compelling Mango's legs to walk along the treadmill.

Bathed in the black light, the monkey's hairy legs appeared

otherworldly on the observation room's monitor. The screen showed a box superimposed at each joint, presenting a constantly shifting numerical value as the animal's legs flexed and extended. The information was being sent to the computer via the Kinect, which measured the depth of each joint angle as the pump swung Mango's left leg out in front of him and moved it back into position.

Like Cherry, Mango wore a plastic crown. But unlike Cherry's system, which was tethered by wires, Mango's unit wirelessly transmitted his neural activity to the bank of computers in the next room. It wasn't a particularly elegant setup. The sealable plastic crown was attached to Mango's skull with the same dental cement researchers use to protect the craniotomy area. Bulky though it might have been, however, the plastic crown gave researchers a sterile area to house Mango's telemetry unit. Powered by onboard batteries, the system enabled Mango to move about the research environment without the threat of tangling or damaging the wires.

"We're basically collecting the first true animals that are controlling a BMI wirelessly," Nicolelis said. "The monkey will basically be free of any tethering or restraint or anything." For the moment, however, the wireless system was merely recording Mango's brain activity. The researchers were still trying to habituate him to the leg braces, and they were just beginning to acquaint him with the task of walking on cue. Their plan, eventually, was to paralyze the monkey's legs temporarily, prompting him to use the wireless BCI to control the exoskeleton. But for now the exoskeleton was under full computer control. The treadmill remained stationary as the monkey's brain signals jumped skyward, his legs moving back and forth.

This early prototype was a long way off from a functional brain-controlled exoskeleton, but it was a start. "I'm not treating this as life-or-death," Nicolelis said. "I joke with the Brazilian president that this is the Brazilian moon shot."

•

It's precisely this sort of bravura that has made Nicolelis so controversial in the field. But his showmanship, coupled with cutting-edge BCI work, has also made him one of its most recognizable figures. It's a role he clearly relishes, and Nicolelis has been the driving force behind many of the field's firsts—from his 1999 paper that coined the phrase "brain-machine interface" to his later monkey experiments.

Through it all, however, Nicolelis insists he is more interested in using BCI to shed light on how the brain functions and to question traditional notions of the biological self. "When we created brain-machine interfaces, it was not to create prosthetic devices; the goal was to have a new tool to probe the brain," he said. "We're using the prosthetic work to develop a completely new theory of how the brain works. Nobody's doing that."

For much of the twentieth century, perhaps no two neuroscientists were more influential than David Hubel and Torsten Wiesel, whose groundbreaking research into the visual processing of cats (among other things) won them the Nobel Prize in 1981. Working with both anesthetized and awake animals, the researchers measured the response of individual neurons as they presented the cats with different visual shapes and patterns.

During their first months of experimentation, the scientists failed to elicit a neural response from the visual stimuli. Then one day the hard edge of the slide plate they used to project their patterns slipped across the screen. It was a mistake, but as so often happens in science, the mistake proved decisive. Unlike earlier stimuli, the diagonal line of the slide edge caused the neuron to spark to life. The researchers realized that the cell responded to the visual stimuli of a line falling across the retina. But there was a wrinkle: the line had to be in a particular position and orientation for the cell to fire rapidly. Change the orientation, and the neuron's response diminished. They found that other cells responded more strongly when the line was at different angles, while still others responded to motion, light, or shadow.

Hubel and Wiesel's research served as the foundation for what became the dominant theory of visual perception: namely, the brain builds complex optical scenes by first perceiving simple features, like lines and shadows. The brain channels that basic visual information to higher brain regions, which form increasingly complex patterns, eventually completing the visual stimulus.

Their studies indicated a clear cause-and-effect relationship between external stimuli and evoked neural responses. More recently, however, neuroscientists have argued that this model gives an incomplete accounting of the conscious brain, arguing that the stimulus-response model doesn't take into account the conscious brain's internal state—not only its expectation of a stimulus, but also its evolutionary history.

While some individual cells are undoubtedly in a direct cause-and-effect relationship with incoming stimuli, those cells hardly tell the whole story. And studying them in isolation has pitfalls of its own. "The widespread use of microelectrodes focused experimental research on the behavior of single neurons and the possibility that their individual properties could account for much of what the brain does," wrote the neurophysiologist and historian James T. McIlwain. "As you sit in a darkened laboratory with your attention riveted to the sounds of the audio monitor and probe a neuron's receptive field with a tiny visual stimulus, it is easy to forget that the cell you are listening to is but one of many that are responding to the stimulus."

Nevertheless, many neuroscientists in the twentieth century were confined to studying individual neurons in anesthetized animals: they simply lacked the ability to record from ensembles of neurons, let alone ensembles across various brain regions in awake animals.

As a young medical student, Nicolelis hatched the idea that by implanting multiple electrodes along different regions of a given brain circuit, he could create a physiological map to visualize how information moves along a neural pathway. By recording from different locations along the circuit, he theorized he would be able to add a fourth dimension—time—to his map, charting the shifting neural signals as they move from lower to higher brain regions.

When he approached his Brazilian mentor with the scheme, the elder neuroscientist was adamant. "It is time for you to finish your thesis, leave the laboratory, and go abroad," Nicolelis recounts in *Beyond Boundaries*, his autobiographical account of his research. "What you want to do, neither I, nor anyone else in Brazil, can help you achieve."

Within a year, Nicolelis had been invited to join John Chapin's lab at Hahnemann University in Philadelphia. Like Nicolelis, Chapin was looking to expand his recording abilities. He wanted to move beyond the practice of recording from single neurons, hoping instead to use a novel technique to record from several neural populations at once. Instead of implanting a single rigid wire in the brain, Chapin wanted to use new arrays with as many as sixteen flexible micro-wires. The new arrays not only would enable them to record from multiple neurons but also would be permanently implanted.

The prevailing theory of sensory perception held that tactile sensations were conveyed through mechanoreceptors in the skin by electrical impulses that moved from the peripheral to the central nervous system. Maps of the sensory cortex showed that particular areas of cortex corresponded to specific areas of the body. Known as the homunculus, these maps roughly parallel an animal's physical body, with certain areas of cortex becoming active when animals receive stimulation at a corresponding body part.

This cortical mapping is perhaps most explicit in the snout region of rodents. Recording from single neurons, earlier researchers had found that the area is organized into a well-delineated grid, with clusters of neurons (or "barrels") mirroring the grid-like pattern of individual whiskers on the animal's snout. Researcher after researcher had demonstrated that neurons in these barrels corresponded to particular whiskers: when the whisker received stimulation, neurons in the related barrel would fire like mad. The correspondence wasn't limited to rodents' primary sensory cortex: similar maps existed in the lower, subcortical relay points that formed neural chains between physical whiskers and the sensory cortex. Accordingly, these subcortical neurons would release a series of action potentials when researchers stimulated the appropriate whisker.

Still, most researchers confined themselves to studying individual neurons, solidifying the notion that those neurons were in a direct and exclusive relationship with specific whiskers. The theory held sway for years, becoming only somewhat more complicated in the 1980s when researchers found neurons could also respond to neighboring whiskers.

By the early 1990s, however, Chapin and Nicolelis had set out to simultaneously record multiple individual neurons that projected to different barrels in the sensory cortex. At first, the researchers used the sharpened shafts of Q-tips to mechanically stimulate individual whiskers of lightly anesthetized rats. Unsurprisingly, they found that neurons that were closely associated with specific whiskers released quick bursts of action potentials when they touched the corresponding whisker. Similarly, they confirmed that neurons would also respond, albeit less robustly, when they stimulated nearby whiskers.

Where their study differed from early work, however, was that they looked at multiple neurons over time. The researchers found that instead of simply having a direct causal relationship with a primary whisker, a neuron's receptive field migrated in the milliseconds following the onset of stimulation. The relationship between a neuron and its primary whisker was not set. Rather, it was dynamic through time and space: a neuron became responsive to different areas of the snout at different times during stimulation. These subcortical maps shifted. They reorganized.

To further test the theory, the researchers began anesthetizing small patches of skin on the rodents' snouts, prohibiting those primary whiskers from sending sensory information to their associated neurons. If their theory was correct, the neurons would reorganize, shifting their receptive fields to areas that were receiving stimulation. Sure enough, within a few seconds of numbing a patch of skin, the rats' whisker maps reorganized to reflect the altered sensory reality.

Armed with this information, Nicolelis and Chapin moved to awake animals. By monitoring multiple groups of neurons, they found that individual neurons throughout the sensory system responded to multiple whiskers. The cells weren't confined to individual whiskers and deaf to others. Rather, they shifted in sensitivity depending on the source of the stimulus.

Importantly, later research in awake animals showed that they experienced neural activity in their sensory cortices *before* their whiskers touched anything at all. Researchers found that when an animal anticipated contact, sensory neurons responded as though the whiskers were already being stimulated. By decoding these anticipatory firing patterns, researchers could even predict whether the rats would correctly identify the source of stimulation before it was applied.

It was a radical departure from the earlier cause-and-effect, or feed-forward, model. Far from being a passive organ that merely reacted to external stimuli, the brain actively constructed experience, anticipating sensation and influencing the brain's reception of incoming sensory stimuli. "Hubel-Wiesel cannot be right, because they said that the pathway was only activated when the periphery is stimulated. Uh-uh: the pathway is being activated throughout," said Nicolelis. "That anticipatory activity modulates the responses that are coming from the whiskers."

Nicolelis and Chapin had the data to support their conclusions, but to hear Nicolelis tell it, academic journals resisted publishing their research. Fellow neuroscientists were highly skeptical, even hostile. "It was like going against Jesus Christ, and his disciples were the most violent gang I ever met in my career. They would go around like the bishops of a church, saying this is not part of the canon! This is not the dogma!" said Nicolelis, adding that his "Brazilian genes" helped a lot. "None of the neuroscience from the twentieth century survived. It's all gone. It's not easy, because we're talking about Nobel laureates who were very important, very fundamental, but their work was context dependent. All the feed-forward models of the brain are going to disappear, because they can't handle any of this: feed-forward neural networks cannot explain expectation."

One problem Nicolelis and Chapin had while trying to sway their colleagues was that they were working in the sensory system. Sure, they could present their data, but rats couldn't give an unequivocal outward expression of their sensation. The data showed neural activity, but the animal's behavior? Its experience? A rat couldn't tell you what it felt. It was a matter of interpretation—an interpretation their colleagues could easily dismiss. "People would say, 'Oh, you're deriving this mathematical formula from the recordings, but who guarantees the animal does anything with that information?'" said Nicolelis. "Of course, it's impossible to answer."

They needed a demonstration that was categorical. They needed a research paradigm to show the importance of neural populations and expectations. They needed something they could measure. They needed, in a word, to move to the motor system. "That's when we came up with the brain-machine interface," he said. "Instead of just recording, let's record and throw this to a device. Let's see if the device reproduces what the animal does with its own body."

Recording from forty-six neurons in a rat's motor cortex, Chapin and Nicolelis trained the animal to press a bar and receive a sip of water. As the researchers ran the animal's neural activity through a computer, the rodent eventually realized it didn't need to physically press the bar. It merely had to think about it to receive a reward.

"It was unbelievable. People would shout at us. People would stand up and say it cannot be. Then we started inviting people to come to see the recordings. These big shots—their eyes would pop out," he said. "It was checkmate."

•

Nicolelis soon expanded on that original paper, moving into the more complex realm of the monkey cortex. This early work attracted the interest of DARPA's Alan Rudolph, who in 2002 provided Nicolelis with $26 million, enabling him to produce a series of headline-grabbing demonstrations—linking his monkeys first to robot arms hundreds of miles away and later to a walking robot in Japan.

DARPA awarded Nicolelis a performance award in 2002 as he went head-to-head with Donoghue and Schwartz. But by the time the agency again bestowed the award on him, in 2007, the landscape at DARPA had already shifted. In response to the wars in Iraq and Afghanistan, DARPA had ushered in the Revolutionizing Prosthetics program, replacing Rudolph with Geoffrey Ling, the hard-charging colonel who enlisted other labs to develop upper-limb prosthetics.

DARPA's shift in focus did not bode well for Nicolelis, who watched as his funding began to disappear. "My reward for getting [the performance award] the second time was that they dumped us to build that arm," he said. "It was almost an obsession. Sure, you can build an arm if you throw money at it, but the question is, can you work with that arm? Can you control that arm? And nothing I have seen in the past three to four years has proved that they can."

Nicolelis says that by 2008 he had lost all of his DARPA funding. "We were the best group in the country by several miles," he said. "We were making all this progress. We started the whole business. But then, when the new director came, we could never get our situation solved. We were being bled without knowing why."

But other researchers didn't find Nicolelis's loss of funding so mysterious. Though he was one of the field's earliest and brightest stars, they claimed his work at times lacked rigor, and many groused that the Brazilian's grand proclamations and provocative research paradigms sometimes promised more than his data could support. Rudolph had been a great advocate of Nicolelis's work, but the singularly focused Ling needed measurable results. "He thinks there's a conspiracy against him, but it's like anyone else: you have to write decent grants, and you have to deliver," said Schwartz. "It's not a mystery. He's got to play by the same rules as all of us."

With the money drying up, Nicolelis says his lab was in jeopardy of closing. He had greatly expanded his operations with earlier grant money. Now he needed to find ways to keep it running. "The NIH budget is going down. NSF is disappearing. The only budget surviving is the military," he said. "It is a secret between scientists. It's called post-DARPA death: You go to DARPA, and you get tons of money like they gave to me. But very few people survive the after-DARPA funding, because you can never replace it."

As his fellow researchers began to work on upper-limb prostheses in DARPA's slipstream, Nicolelis was already moving on. To his mind, his monkey work from the early years of the twenty-first century had demonstrated all there was to show that upper-limb BCIs were possible. "People are afraid of risking it. People play safe. The system encourages that," he said. "Just little tiny details and boom! It's a *Nature* paper. We think totally different. We want to describe the big macro picture. Not just for upper limbs, but for everything—locomotion, new sensory signals, because I think that's where we're going to have the real big prize."

•

To that end, Nicolelis's rat lab boasted row after row of amber-colored plastic cages. The containers, each about the size of a shoe box, housed a few score of black mice. Some were balled up and resting. Others moved freely. But these were no regular mice. Each mouse sported an implant atop its skull. They were also transgenic, genetically engineered to exhibit certain disease traits like Parkinson's or OCD-like symptoms.

Like the monkey lab, Nicolelis's rodent lab was a large enterprise—several rooms filled with old monitors, beakers, vials, and powdered chemicals. To the left was a small operating room where researchers used microscopes to install electrodes in mice and rats. To the right were several large rooms brimming with electrode leads hanging from the wall and silver ventilation tubes on the ceiling. In the middle of one of the lab's central rooms, researchers had set up a "behavior chamber" where the researcher Eric Thomson was working with a rat named Teal.

Thomson is a tall man. He parts his brown hair to the side and wears black wire-rimmed glasses. He would move excitedly through the lab, grabbing this paper or that study, hunting down a specific electrode, and marveling at the smallness of some of his implants. His setup wasn't as elaborate as the monkey lab, but that was okay. He is, in his own words, "more of a rat guy" anyway.

The focus of his attention these days, though, was the behavior chamber, a large square aquarium they'd draped in black cloth. Researchers had placed a black cylinder inside the cube. The cylinder was a sort of rodent arena. At its base were three nodes that formed a triangle. The nodes looked something like stoplights, but instead of telling the animal to stop or go, the nodes' "red light" emitted an infrared light, the "yellow light" released a water reward, and the "green light" emitted a traditional light.

To the left of this contraption was a blue carrel that housed a small monitor showing Teal in the darkened chamber. Nearby, a computer was connected to a lab-made neural stimulator. The green circuit board, about the size of a VHS tape, hosted a multitude of transistors and wires, which sent small electrical pulses to Teal's sensory cortex.

The task here was to endow Teal with infrared "vision." The rat had an infrared sensor attached to its head, and the idea was that the sensor would register each time one of the chamber's infrared lights became active. The sensor would speak to the computer, which would send brief pulses of electricity to the animal's sensory cortex. The stimulation frequency increased as Teal approached the light source, enabling the animal to gauge how far it was from the light, or, in rat terms, from its water reward.

Researchers had used both traditional and infrared light to train the rodent. When the lights shone, they stimulated Teal's brain, increasing the frequency as the animal approached the lights. Once Teal made the connection between increased stimulation frequency and the reward, they turned off the normal light. Left with only infrared-based stimulation, the rat soon learned to associate increased stimulation alone with a water reward. The animal couldn't see the infrared light source, but outfitted with an implant and infrared sensor, it could nevertheless locate the active node, endowing Teal with what Nicolelis called a sixth sense. "It doesn't see the light, it feels a tactile stimulus, but what emerges from that?" he asked. "We don't know."

One result that Nicolelis found particularly interesting was that Teal's behavior changed over time. Whereas the rat had initially pawed at her snout when researchers stimulated her sensory cortex, she eventually gave up that behavior. Instead, she began moving her head back and forth, scanning the chamber as she sought out the infrared light. "Not only did the animal perceive the infrared, but that influenced the whole behavior of the animal. An animal that used to walk straight like any rat does, now walks with this sweeping motion," he said. "It sounds like a simple thing, but it altered the animal's entire behavior to find the water."

Sure enough, as Teal appeared on the lab's small black-and-white monitor, she turned left and right, hoping to perceive an infrared signal. The camera they'd mounted atop the cylinder made the chamber's base look like a mottled moonscape, as Teal, her head tethered to a mesh cable, received information from her extended nervous system, silently processing this novel sense of vision. As she swept the area, one of the nodes lit up. The animal's oscillatory movements quickly diminished as she homed in on the light source, racing in a straight line to her water reward.

In a characteristic flourish, Nicolelis had opted to stimulate the rat's sensory cortex (as opposed to the visual cortex). In the classic conception of the brain, specific areas like the sensory and visual cortices were thought to be linked to specific functions, like touch or vision. Scientists believed neurons were devoted solely to those functions and incapable of taking on new modalities. It wasn't until researchers discovered the principles of neuroplasticity that they began to theorize that areas of cortex normally associated with one function might be recruited to others. "What we're finding with the BCI and other work is that it's more of a continuum. These borders that we defined don't make much sense for the brain," Nicolelis said. "We can induce a piece of cortex that is theoretically related to touch to process information about a completely different modality, like infrared."

Neurons formerly associated with one sensory input could be harnessed to process other senses, enabling Teal to perceive a portion of the light spectrum that very few mammals have evolved to see on their own. "We are transforming infrared perception," he said. "It's almost like the guy is touching it, but it's not touching the body anymore. It's out there in the world."

Still, Teal wasn't able to say what she had perceived, and it wasn't entirely clear that the rat was experiencing something we'd recognize as "vision." Certainly, the animal responded to stimulation of its sensory cortex, but the study left open the question of whether Teal was responding to a visual depiction of infrared light or if she was merely responding to increased stimulation. The stimulus was linked to infrared light, but that could be incidental: the rat's experience might have been purely physical.

But Nicolelis insists something else is at play. "What we found is that the neurons basically start responding to both touch and infrared. The infrared does not hijack the cortex," he said. "We are making them feel light."

•

Perhaps even more radical was the vein of research Nicolelis and his collaborators were pursuing to create a BCI that linked the brains of two animals. Working once again with implanted rats, the researchers placed the animals in separate behavior chambers. A wall in each chamber was outfitted with a pair of levers—one lever to the right, another to the left. Above each lever was a light. The chamber's opposite wall housed a funnel where the animals received their juice reward.

The task was fairly simple. Researchers trained the first rat, known as the encoder, to press a lever each time its corresponding light came on. The animal received a juice reward whenever it chose correctly. With a simple task like this, a classic, really, the encoder rat had a near-perfect success rate.

Meanwhile, researchers used electrodes to record the animal's brain activity, mining it for specific features linked to deciding which bar to press and the action of pressing the bar. Where the experiment differed, however, was what happened next. Taking the encoding animal's recordings, they transferred the neural pattern to a separate computer, which in turn "uploaded" the stimulation pattern to a second rat, known as the decoder. The second animal did not have the same visual cues as the first rat. Both lights in its behavior chamber remained on. The decoder rat had to rely exclusively on neural stimulation to decide which bar was the correct one to press to receive its reward.

Previous researchers had shown they could control animal behavior through micro-stimulation, prompting rats to move right, left, and forward by sending specific signals to the brain. Nicolelis's decoding rat, by contrast, was not only receiving biologically generated stimulation patterns but actually making sense of those patterns, using them to guide its own behavior. "The rat has no idea what it's supposed to do," said Nicolelis. "But he's able to decode the brain activity that comes from the first rat and reproduce the behavior." In other words, the decoding rat was having a similar neural experience as the encoding rat.

To make things more interesting, Nicolelis then closed the loop, linking the first rat's reward to the second rat's performance. In the experimental paradigm, the first rat received a juice reward only if the second rat successfully completed the task. Because the first animal had already successfully completed the task, it expected its reward. What the researchers found, however, was if that reward wasn't forthcoming, the first rat would concentrate more intensely during the next trial, enhancing its brain signal and making it more readable to the second animal. "They are actually working together," Nicolelis said. "We made one brain out of two brains. It's a super brain—an organic computer."

In a set of related experiments, Nicolelis and his colleagues stimulated the whiskers of the first rat by having it explore various-sized apertures. If the aperture was small, the animal received a reward if it moved to the left. If the aperture was large, the animal was rewarded when it moved to the right. After uploading the first animal's neural activity to the second rat, they again closed the loop, linking the first animal's reward to the second animal's performance. Researchers found that the second animal could correctly read the incoming signals 65 percent of the time.

For Nicolelis, it's this sort of brain-to-brain interface that not only sets him apart from other researchers but also suggests the true potential of neuroprosthetics. "It's not about moving an arm. It's about suggesting that the brain is so plastic that it can incorporate another body as its source of information to probe the world," he said. "That touches on theories of self, theories of identity. Once you connect brains like that, who is to say there's not another level of emergent properties that materialize by the interaction of the two brains?"

It's just this sort of speculation that makes many of Nicolelis's rivals bristle. "We are all interested in these very provocative questions," said Brown University's John Donoghue when asked about Nicolelis's work. "We are all deadly serious scientists who are interested in how the brain works." Nicolelis, he said by way of contrast, seems less interested in basic science. "Is this for showmanship? Why is it done? I don't know. The roots of all the things have already been done. If it's a vehicle for provoking conversation, it's sort of not how we usually do that in science," he said. "It's another niche. It seems to get more and more marginalized and less and less interesting—except for the press."

But Nicolelis is undeterred by his old rival's criticism. He insists that he is more interested in unlocking the mysteries of the brain and consciousness than in merely making incremental progress on an already established proof of principle.

"What I'm trying to see is if you put several brains to work like this, you may have a result we cannot even predict. We may be able to compute things that a single brain could not compute," he said. "None of the BCI literature on upper limb outside this lab touches on that. It's not that they're not doing it. It's worse than that: nobody's even thinking about it."

8. CYBERKINETICS

In the summer of 2012, neuroprosthetists from around the globe traveled to Salt Lake City, where for three days they combed through some of BCI's most stubborn questions: How could they minimize the body's reaction to their devices? Are there better ways to merge electrodes with neurons? How far away are we from having a fully implantable wireless device?

The tools of their labor were everywhere on hand at the Salt Palace Convention Center, where vendors had set up displays showcasing their latest devices. The neurotech firm Blackrock Microsystems displayed several mannequins depicting the digitally integrated brain. One model had an implant just behind the right ear, its thin gold wire exiting the skull via a port atop the cranium. Nearby, a cross section of preserved cortex had a similar array. But perhaps the company's most impressive display was a model of a human torso and skull, a commercial vision of today's augmented self. The model carried an implanted power source just below the clavicle that was about the size of a dental floss case. From this case flowed wires that extended up the neck and connected to three electrode arrays, each no larger than a Tic Tac, implanted in the model's brain. Nearby, an EEG device manufacturer named Brain Products had outfitted an eerily realistic baby doll with a tightly fitting electrode cap that sprouted a Medusa's wig of electrical leads.

But the largest booth by far belonged to DARPA, which had set up a freestanding temporary office. Constructed of blue panels emblazoned with the DARPA logo, the makeshift office displayed placards of the various projects the agency has funded. An unfailingly polite assistant sat at a simple desk outside the office, manning a laptop computer while scheduling ten-minute meetings between scientists and Jack Judy, the agency's preppy manager for the RE-NET program, which seeks to increase the durability of implants. DARPA was one of the conference's main funders, and throughout the weekend Judy would disappear with investigators behind his makeshift walls to hear their pitches, hoping they would match his focus area.

"Ultimately, we'd like to make devices that don't exhibit degradations," Judy told the assembled scientists by way of introduction. "I'm from DARPA. I care about amputees. I care about pain. I care about other things like that. I don't really care about Parkinson's disease." As with Ling and the Revolutionizing Prosthetics program, Judy was less focused on patients with spinal cord injury or who suffered locked-in syndrome. Rather, the DARPA program manager was interested in rehabilitating soldiers. "Young individuals," he said, "who are athletic, and who don't want to have a geriatric limb attached to their bodies. The demands are quite high."

•

Still, the stars of the conference were John Donoghue and Leigh Hochberg, careful, diplomatic brain scientists who earlier that month had made international headlines when they granted quadriplegics direct neural control over a prosthetic limb. Unlike the higher-ups at DARPA, who aimed to rehabilitate athletic young vets, Donoghue and Hochberg worked with an unspeakably vulnerable patient population: not merely quadriplegics, but locked-ins, people who have lost their ability to eat, breathe, and even speak on their own. Locked-in syndrome is often the terminal stage of amyotrophic lateral sclerosis, or Lou Gehrig's disease, where the body's motor neurons die off, eventually leaving the victim paralyzed from head to toe, his consciousness locked inside an immobile body.

The British historian Tony Judt, who suffered from ALS before his 2010 death, once explained that the disease's progressive paralysis was like being in a prison cell that shrinks by the day. "You don't know when it's going to get so small it's going to crush you to death. But you do know it's going to happen, the only question is when." In his essay "Night," Judt wrote movingly that "having no use of my arms, I cannot scratch an itch, adjust my spectacles, remove food particles from my teeth, or anything else that—as a moment's reflection will confirm—we all do dozens of times a day."

During the day, Judt, who relied on a ventilator and lost control of his limbs while retaining sensation, could ask people to adjust an arm or shift his body. "But then comes the night," he wrote, when the able-bodied slept and he was left alone, trapped and immobile. "Every muscle felt in need of movement, every inch of skin itched, my bladder found mysterious ways to refill itself in the night and thus require relief, and in general I felt a desperate need for the reassurance of light, company, and the simple comforts of human intercourse." Mornings brought the meager promise of his wheelchair, human contact, and a welcome shift of his deadened limbs, but it also brought the understanding that his life and abilities were in a state of incremental decline—one that would inevitably lead to the loss of his voice, voluntary muscle control, even the ability to swallow. "I wake up in exactly the position, frame of mind, and state of suspended despair with which I went to bed—which in the circumstances might be thought a considerable achievement."

Like Judt, the two participants in Donoghue and Hochberg's study were locked in. Unlike Judt, however, they had arrived at this pitiable state suddenly, suffering brain stem strokes that knocked out a critical link between mind and body. As its name implies, the brain stem is located at the base of the brain, a collection of ancient neural structures that includes the medulla oblongata, the pons, and the midbrain. This neural archipelago has a role in everything from transmitting signals from the forebrain to the cerebellum, to regulating essential bodily functions like sleep, breathing, swallowing, and blood pressure. The slender channel also acts as an indispensable passageway for the body's motor neurons (indeed, "pons" is Latin for "bridge"), linking the brain with the spinal cord and peripheral nervous system.

A brain stem stroke disables this mind-body link, leaving the victim's brain fully intact but incapable of communicating with the body. And that's the real calamity: victims are unable to make most physical gestures, but their cognitive function remains fully intact. They are totally conscious, but their consciousness is imprisoned, locked inside the cell of their inanimate body.

The voices of those locked in are astonishingly rare. Perhaps the best-known account came from Jean-Dominique Bauby, the former editor of French *Elle* who suffered a brain stem stroke at the age of forty-three. "I had never even heard of the brain stem," Bauby, who communicated by blinking his left eye, wrote in his memoir, *The Diving Bell and the Butterfly*. "In the past, it was known as a 'massive stroke,' and you simply died. But improved resuscitation techniques have now prolonged and refined the agony."

On that day in December 1995, Bauby had awoken, "heedless, perhaps a little grumpy." He spent some time at the office and suffered through a business lunch before picking up his son for the weekend. He had planned to go to the theater that evening, but Bauby's body began to move in slow motion as he arrived to collect his son. He was having trouble driving the car, and he began to see double as sweat formed on his brow. "I stagger from the BMW, almost unable to stand upright, and collapse on the rear seat."

Bauby was suffering a cerebrovascular accident at the base of his skull, cutting off the blood supply to his brain stem and causing massive neuron death. Bauby tried to tell the driver to slow down as they raced to the hospital, but he'd already lost control of his mouth and vocal cords. "No sound comes from my mouth, and my head, no longer under my control, wobbles on my neck." When they finally arrived, Bauby had time for one last thought, "We'll have to cancel the play," before he sank into a coma lasting three weeks.

Like Bauby, the study's participants were seemingly in good health when they suffered their strokes. The first, a fifty-three-year-old mother of two known as S3 in the scientific literature but as Cathy Hutchinson to her friends, had been planting vegetables in the spring of 1996 when she suddenly became nauseated. As the journalist Jessica Benko recounts, Hutchinson's ears filled with buzzing, and she was soon unconscious. It took doctors roughly twelve hours to discern that she had suffered a brain stem stroke, but by then it was too late: the damage was done. Hutchinson slid into a deep coma.

Hutchinson emerged three weeks later only to realize that no matter how hard she tried, she could not lift her hand. She couldn't move her legs or form words with her mouth (let alone expel air across her larynx). "Of course, the party chiefly concerned is the last to hear the good news," Bauby wrote of waking to find himself trapped in his flesh-and-bone sarcophagus.

Bauby called the prison of his body his diving bell, where he spent eighteen months before finally succumbing to pneumonia in 1997, a mere three days after publishing his memoir. His mind, by contrast, was his butterfly. It enabled him to escape his confinement, revisiting past meals, lovers, trips, and fantasies. "You can visit the woman you love, slide down beside her and stroke her still-sleeping face. You can build castles in Spain, steal the Golden Fleece, discover Atlantis, realize your childhood dreams and adult ambitions." But then he would return to his diving bell, reaching out to the world with only his left eye.

Bauby used that eye to spell out his memoir one letter at a time, blinking in affirmation each time his speech therapist, reciting the alphabet, arrived at Bauby's intended letter. The therapist would then repeat the drill, reading the alphabet until words, sentences, paragraphs, and ultimately a memoir finally emerged. "In my head I churn over every sentence ten times, delete a word, add an adjective, and learn my text by heart," was how Bauby described preparing for his writing sessions. It was slow going, making banter with loved ones difficult.

With such a slender tether to the outside world, Bauby was given to making wry observations about life inside his diving bell, as when he didn't recognize his own reflection. "I saw the head of a man who seemed to have emerged from a vat of formaldehyde. His mouth was twisted, his nose damaged, his hair tousled, his gaze full of fear," he wrote. "One eye was sewn shut, the other goggled like the doomed eye of Cain. For a moment I stared at that dilated pupil, before I realized it was only mine."

That doomed eye was Bauby's only portal to the outside world, and the thought of losing it, as when he awoke one morning as the hospital ophthalmologist sewed shut his malfunctioning right eye "as if he were darning a sock," engulfed him in fear. "What if this man got carried away and sewed up my left eye as well, my only link to the outside world, the only window to my cell, the one tiny opening of my diving bell?"

•

Given to gray slacks and blue blazers, John Donoghue cuts an avuncular figure with wispy white hair and beard to match. He walks gingerly on an arthritic hip, a result of a childhood case of Perthes disease, a temporary loss of blood to the femur that causes the bone to die. In Donoghue's case, the bone died from the inside out, and doctors didn't discover he had the disease until the ball of his femur collapsed.

Donoghue's bones eventually healed, but he spent a year in bed, eventually graduating to a wheelchair and leg brace before regaining full mobility. His convalescence taught him to appreciate how devastating motor disease could be. He knew firsthand what it was like to be dependent, incapable of running or riding a bike or many of the other things his friends didn't think twice about. "These are things that probably have a big impact, but I wasn't saying, okay, at that moment, I'm going to cure motor disease," he said. "It would be a great story, but it's not true."

Instead, Donoghue went on to study biology at Boston University, later working for the neuroanatomist Paul Yakovlev. He studied anatomy at the University of Vermont before heading to Brown University to earn a doctorate in neuroscience. While at Brown, Donoghue became fascinated by the emerging field of neuroplasticity. He worked briefly with the rhythmic firing of groups of neurons, known as neural oscillations, while also concentrating on the motor cortex, investigating first how the brain represents different areas of the arm and later how it behaves during voluntary movement.

In the early 1990s, however, he met Richard Normann, a researcher at the University of Utah. Normann was interested in visual neuroprosthetics, and he was developing novel ways to stimulate the visual cortex with microelectrode arrays. Normann's array, which became known as the Utah array, was ingenious in its design. The brain can travel up to two millimeters inside the skull. Earlier electrodes had fastened to an immobile pedestal that was attached to the cranium. The Utah array, by contrast, anchored to the brain itself, enabling it to "float" with the brain, keeping close contact with targeted neurons as they shifted in the brain case. What's more, instead of piercing or damaging individual cells, the Utah array merely displaced neurons, leaving them biologically intact for recording.

Normann was using his array to stimulate the visual cortex, delivering small pulses of electricity to approximate vision. But Donoghue wanted to use the array in the opposite manner, implanting it to record nearby action potentials he could feed into a computer.

By 2000, he and his colleagues had procured a $4.25 million grant from the Department of Defense to explore the brain using

Normann's array. Two years later, they published a brief paper in *Nature* demonstrating that with the Utah array they could grant monkeys instant neural control over a cursor.

Nicolelis had already made a splash with his groundbreaking owl monkey experiments. Philip Kennedy, meanwhile, had implanted John Ray with his neurotrophic electrode. What set Donoghue's paper apart was the Utah array, which the Brown group argued stayed in closer contact with neurons than earlier electrodes. The array's microelectrodes also enabled researchers to record from ensembles of neurons without the clutter of individual micro-wires. Better yet, its platinum tips and silicon body could better tolerate the harsh environment of the brain, delivering stable recordings for longer periods of time.

The Utah array, Donoghue argued, was the sort of implant that could be suitable for human use.

With the exception of Kennedy, invasive researchers were working exclusively in animals. DARPA's Revolutionizing Prosthetics program was still years away, and Donoghue realized it might take decades for academic labs to make the leap to human clinical trials. With their focus on basic science over applied results, university labs simply lacked the resources, discipline, and incentive to develop a neuroprosthetic for human use. Not only would the technical hurdles of redesigning the Utah array for humans be immense, but they would also have to demonstrate it was safe—an onerous bureaucratic task the FDA required before it would approve a human study. "That would eventually happen in academics, but it would have taken many, many, many more years," said Donoghue. "Look at all the labs that have tried to do it on their own. You can argue that no one has succeeded."

•

Donoghue believed that the only way to open the human cranium to a computer interface would be through a private company, one with the capital and focus to reengineer the array and punch through the FDA's thicket of bureaucracy. Working with a small group of fellow researchers, Donoghue cofounded Cyberkinetics Inc. in 2001. Its aim was to develop the BrainGate, an assistive BCI to give quadriplegics and other motor-impaired consumers neural control of computers and prosthetic limbs.

The group quickly secured investors, raising more than $9 million by 2003. Meanwhile, Cyberkinetics merged with Bionic Technologies, a neurotech firm cofounded by Normann, who held the patent for the Utah array. As they lined up their intellectual property, Cyberkinetics added a neural decoding patent out of Brown, while also licensing an astonishingly broad patent held by Emory University's Donald Humphrey. Granted in 2001, Humphrey's patent laid claim to any system that uses sensors implanted in the central nervous system to record and process neural signals that are transmitted to an external device. In other words, most any BCI.

As they worked to secure funding and intellectual property, the Cyberkinetics team was also busy modifying the Utah array, improving the company's surgical techniques and software. The Utah array was foundational to the BrainGate system and improving its functioning, implantation, and overall safety dominated the FDA process. The team, which also included Mijail Serruya, Gerhard Friehs, and Nicholas Hatsopoulos, had implanted dozens of arrays over the years. They'd worked with nearly twenty monkeys, some for as long as thirty-three months. In advance of their FDA application, however, they demonstrated that the redesigned Utah array could be safely implanted for more than a year in three macaques.

Cyberkinetics was moving forward on other fronts as well, recruiting human subjects, working out contracts, and gaining approval from Institutional Review Boards at the proposed trial sites. The FDA eventually signed off on a pilot clinical study to implant human research subjects with the BrainGate system.

With FDA approval, Cyberkinetics now seemed poised to push the field dramatically forward, and they quickly arranged interviews with a few choice media outlets. "You can substitute brain control for hand control, basically," Donoghue triumphantly told

The New York Times. In the same article, the Cyberkinetics CEO, Timothy Surgenor, estimated they'd have a marketable device by "2007 or 2008," and executives estimated the system would enable users to type "as fast as a healthy person could type on a BlackBerry."

Still, people in the field were skeptical. Dawn Taylor, a researcher at Case Western Reserve University in Cleveland, told the paper, "A disaster at this early stage could set the whole field back." Kennedy, the only other researcher to work in humans, worried about the system's percutaneous wiring, noting, "We don't like to hang around with wires coming out of our head."

Cyberkinetics required that its study participants be paralyzed but able to speak, prompting Nicolelis to question the very need for neural cursor control when other assistive technologies like eye tracking and speech recognition programs were already available. "If you are only talking about moving a cursor up and down on the screen, you don't need to get into the brain to do that," he told the paper.

Nicolelis had spent much of the previous decade doing multielectrode neural recordings. His technique was not unlike the system described in the Humphrey patent Donoghue intended to use. But by moving so decisively into humans, Donoghue had seized the spotlight, prompting Nicolelis to accuse Cyberkinetics of co-opting ideas that were already in the public realm to start the company.

"He took advantage of a lot of stuff that was published and tried to take it for profit very quickly," Nicolelis said. "Nobody would be doing BCI if it were not for the capacity of doing chronic multielectrode recordings. That's what John Chapin and I were doing when I was a postdoc in his lab."

•

By June 2004, the Cyberkinetics team had recruited the former footballer Matthew Nagle. Later that month, Friehs used a pneumatic wand to inject a Utah array into Nagle's motor cortex, installing a pedestal atop his skull to connect him to the BrainGate system. "What took a year at Cyberkinetics would've taken a decade in an academic setting," said Donoghue. "In order to get an FDA filing, in order to get all of the information that is needed to say is this safe electrically, is it safe in terms of biocompatibility—all of that stuff had to be overseen by people who understood the right way to do it. Cyberkinetics did all of that."

Researchers started Nagle with basic center-out tasks, moving a computer cursor to one of several peripheral targets. He soon graduated to more complicated tasks, using the BrainGate to play *Pong*, change the channel on a television, and navigate a computer desktop.

Still, as a biomedical start-up with only $10 million in funding, Cyberkinetics was woefully underfunded. It needed to show results. It needed to garner good press and raise more capital. The privately held company was also about to go public via a reverse merger with the defunct Trafalgar Ventures, a publicly traded Canadian mining company. Absorbing the shuttered mining operation would give Cyberkinetics access to the public market. It was a cheap way to create a public company, and a splashy demonstration of Nagle's progress would help drum up interest for its initial public offering. "It was a bid to find another way to capitalize the company," said Donoghue, who presented the team's preliminary data at the American Academy of Physical Medicine and Rehabilitation meeting.

In the days following the IPO, Cyberkinetics' stock price tripled to $6.50, as the company placed several positive stories in national media. "It's Luke Skywalker," Donoghue told *USA Today*. Friehs described Nagle's results to CNN as "spectacular" and "almost unbelievable," adding, "We have a research participant who is capable of controlling his environment by thought alone—something we have only found in science fiction so far."

Flush with their successful IPO and publicity campaign, company executives began circulating optimistic e-mails, detailing the firm's expectation of having a marketable product within three years. As Jon Mukand, the clinical investigator on the study, recounts in his book *The Man with the Bionic Brain*, Vice President Burke Barrett wrote that he expected research technicians soon would no longer be necessary, because the system "could be used by the patient every day with the help of a caregiver." Barrett added that for 2006, Cyberkinetics planned to introduce a wireless device that communicated with a "smaller computer that was powered off the wheelchair," imagining "advanced versions that would use/link with robotics, computerized muscle stimulation, etc."

•

Schwartz watched with trepidation. He was still waiting in the wings at DARPA, but between Donoghue's success and his own work with monkeys, administrators at the University of Pittsburgh Medical Center approached him about forming his own company, assigning him a market analyst. "We came to the conclusion that it was a really bad idea. The market wasn't there. The technology wasn't mature," Schwartz said. "This was in the heyday of Cyberkinetics, and this analyst said, 'They're going to fail. They're going to fail soon.' "

Privately, Donoghue and his colleagues were coming to the same conclusion. They still hadn't attracted a second research subject. More important, money was running low. The company had incurred losses every quarter since its inception, and its chances of delivering a marketable product seemed increasingly dubious. "The extent of our future operating losses and the timing of profitability are highly uncertain, and we may never achieve or sustain profitability," the company admitted in a 2005 filing with the Securities and Exchange Commission, adding that it had accrued $17 million in debt. "We anticipate that we will continue to incur operating losses for the foreseeable future and it is possible that we will never generate substantial revenues." Surgenor also noted that Cyberkinetics had yet to demonstrate its device was safe for chronic human use or that it could obtain the "regulatory approvals necessary to commercialize products."

Nevertheless, the public laurels continued. Toward the end of 2004, *Discover* magazine presented Donoghue with its Innovation Award for Neuroscience. The American Institute for Medical and Biological Engineering named him a fellow in 2005, the same year *Wired* magazine chose the BrainGate as one of its top scientific or technical discoveries.

Nagle, meanwhile, was also making the media rounds, appearing in everything from local papers to Germany's *Der Spiegel.* "I can bring the cursor just about anywhere," Nagle told *Wired* in March 2005. "When I first realized I could control it I said, 'Holy shit! I like this.'" Nagle went on to say he was convinced the BrainGate would soon restore his movement. "It's just around the corner," he said. "I know I'm going to beat this." And he later added, "I can stick with it another two years, till they get this thing perfected."

Privately, though, Nagle was growing impatient. Mukand recounts that Nagle had always believed he would move again, once telling a friend that doctors could amputate his arm and replace it with a neurally controlled robotic limb. But as the study progressed, Nagle became increasingly interested in other treatments. He looked into functional electrical stimulation (FES) systems, which paraplegics have used to transfer out of wheelchairs and walk short distances. The systems have also been harnessed to animate arms paralyzed by stroke, enabling users to open jars and other actions that require generalized movement patterns.

One of the long-term goals at Cyberkinetics was to integrate its brain implants with an FES system. But that was a long way off. And for a quad like Nagle, an FES-integrated neuroprosthetic would not merely have to animate the hand. It would also need to control the wrist, elbow, and shoulder—a daunting challenge for a system that had managed only three degrees of freedom on a computer screen.

But even if they had the technical know-how to marry their system to FES, the study's protocols barred Nagle from receiving any other implants as long as he harbored the BrainGate. What's more, Nagle's disability was so profound that it was unclear if he would benefit from such a system. "I did not believe he would benefit much from functional electrical stimulation," wrote Mukand. "But I held his paralyzed and insensate hand, nodded in agreement, and forced out a smile that said it was only a matter of time."

Meanwhile, Nagle was plagued with many of the side effects of paralysis. A series of urinary tract infections kept him on a steady regimen of antibiotics. Spasticity in his legs interrupted his sleep patterns. His trachea was damaged and raw from his ventilator, and after years in a wheelchair he had developed type 2 diabetes, which only increased his chances of skin and brain infections associated with the BrainGate.

The ventilator was a constant irritant for Nagle, who longed to be free from its chafing tubes. Unable to breathe on his own, he wanted to have a phrenic pacemaker, an implanted system that uses electrodes to stimulate the diaphragm and fill the lungs with oxygen. But again, the Cyberkinetics protocol prohibited the device.

As the study progressed, Nagle was also embroiled in the trial of his attacker, Nicholas Cirignano. Strapped to his chair, he delivered his testimony in a scratchy whisper: "You got me that night, but I tell you, you won't beat me. I'm not going to live my life as a loser like you." He later added, "I can't believe I am sitting here in this chair. I look out the window and say, 'This is my life.' "

The Norfolk Superior Court eventually found Cirignano guilty, sentencing him to nearly a decade in prison.

•

With the yearlong study drawing to a close, the Cyberkinetics team hoped Nagle would opt to retain the implant. But by then, only half of the BrainGate's electrodes were delivering signals.

Nagle's diabetes had increased his risk of infection. He yearned to breathe without the ventilator's mechanical rasp, and in October 2005 he underwent brain surgery to remove the implant.

Nevertheless, Nagle was a pioneer. *Nature* fast-tracked Donoghue's article, placing Nagle on the cover of its July 2006 issue with an accompanying editorial that asked, "Is this the bionic man?"

Donoghue's colleagues, however, were dismayed by what they saw as the study's scant scientific information. "The performance of the cursor he had? It was equivalent to EEG," said Nicolelis. "Scientifically, there was no contribution. He gives talks that look like a company public relations guy. He's showing videos. He's showing clips. Where are the recordings? What happens to the brains of those patients? Did the tuning properties change? There is no science behind it."

Particularly galling, the company's rivals said, was a video the Cyberkinetics group circulated on day 114 of the trial. The clip shows Nagle in a blue T-shirt with a thick cable budding *Matrix*-like from his head. A disembodied prosthetic hand rests before him on a burgundy cloth. As Nagle says "close," the hand's thumb and forefinger press together in a basic pinching action. "Holy shit!" he exclaims. "Nice!" After making a few more pinching actions, he marvels, "Not bad, man, not bad at all."

Schwartz, on the other hand, disagreed. "They were so transparent. You know the 'holy shit' thing? That's binary control! You know, one neuron: on, off," he said. "In a way, I was relieved because the performance was so bad. If they had had really good performance, I'd be completely screwed."

One thing Donoghue's colleagues objected to was that by publishing the company's research in an academic journal, the scientists were not as forthcoming about their methods as those who don't have a financial stake in their findings. "There's always that thing," Schwartz said. "You ask them a question, and they say, 'I can't answer that. My lawyers won't let me answer that.' "

A few years prior to publishing the BrainGate study, the editors at *Nature* had published an editorial titled "Is the University-Industrial Complex Out of Control?" The editorial focused on biomedicine, but it raised a related issue, noting that "researchers sponsored by companies are biased in favour of reporting positive experimental results relating to company products" while down-playing, or even omitting, results that might be scientifically relevant but bad for business.

But while the BrainGate study was undoubtedly a triumph, the Cyberkinetics partners continued to worry about the business. The system's complicated interface required a specialist (preferably with a PhD) to calibrate it each day and run the software. What's more, the device would never be marketable so long as it required a dishwasher-sized bank of computers that wired percutaneously into the brain.

Researchers conceded these difficulties toward the end of their *Nature* article, writing that the system leashes users "to a bulky cart and requires operation by a trained technician. A wireless, implantable, and miniaturized system combined with automation will be required for practical use." But these were technical issues, and Donoghue believed understanding the brain was the real challenge. The technology? That was just a matter of engineering. "Emerging and available technologies appear to be sufficient to overcome these obstacles," they continued, "although the challenges of creating a fully implantable system may be formidable."

Eight years after publishing the article, Donoghue was more circumspect. "I thought that within five years we would have a wireless system implanted in people," he said. "I wasn't an engineer. I wasn't able to appreciate how incredibly complex it was."

Still, an even larger problem remained: the intended market wasn't large enough to deliver a reasonable return to investors. "Think about it," said Schwartz. "What was the market? The possible market was spinal cord." But not just any spinal cord injury. Someone who retained use of her arms had no need for a brain implant to control a cursor. "You have to be C4 and above to get any kind of benefit out of what they had," he said. "So what's your market? Maybe a few hundred people? And how many of those are going to go for a brain implant?"

As Leuthardt would later discover with ECoG, the math for spinal cord injury was daunting. It would take millions to develop the device, and even then Cyberkinetics estimated its intended market would remain in the "single-digit thousands." "You burn through all this cash, and you sell your first device, but that's not making a profit," said Schwartz. "So at what point are you going to start making money?"

•

Nevertheless, Cyberkinetics continued to recruit new patients, most notably Cathy Hutchinson in 2006. Hutchinson had terrific signals at first, but around her two hundredth day in the trial she was dropped while being transferred. The Utah array was designed to move with the brain, but the surgeon had inadvertently anchored it with a suture to Hutchinson's dura mater. When Hutchinson fell, her brain shifted but the electrode remained in place, dislocating from its intended cells. "How is something that has to have ten- to twenty-micron precision going to be yanked millimeters?" Donoghue said. Hutchinson's neural count dropped precipitously. "From then on, it was unstable. It stayed anywhere from some low number up to in the forties or fifties."

Their other research subjects were plagued with even worse problems. Stephen Heywood, an architectural designer who suffered from advanced ALS, died in late 2006 of unrelated respiratory failure while still in the study. Nagle, meanwhile, underwent thoracic surgery after the study to implant a phrenic pacemaker. He later succumbed to an unrelated case of sepsis, slipping into a coma before dying in July 2007.

Between the uncertainty of the eventual market and the state of the technology, Cyberkinetics realized it needed to shift its focus and start generating cash. "The business people realized we weren't going to get to a point of making money unless we did something else," Donoghue said. What they hit upon was Andara

Life Science Inc., an Indiana-based company that was developing a technology to regenerate damaged nerves.

By the time Cyberkinetics acquired Andara through a merger, the Indiana-based company had already conducted several animal studies where researchers applied oscillating voltages across the injured area of the spinal cord, which caused some of the nerve fibers to regenerate. Andara followed up these early studies with a small clinical trial, showing that the treatment could partially reestablish some neural channels in a subset of nerves.

"It wasn't great, but it had promise," said Donoghue. With these results in hand, the company appealed to the FDA for a humanitarian device exemption, which would expedite the regulatory approval process. The neural stimulation market was estimated at some $1.6 billion. Andara seemed closer to market than the BrainGate, so Cyberkinetics back-burnered BCI, hoping the Andara would sustain the company until the BrainGate was ready.

The company needed to generate income of its own, and quickly. But even if everything went smoothly with Andara, it would be a few years before Donoghue and his partners could expect a product, and even then there was no guarantee it would sell. "We didn't realize the magnitude of the cost of doing this until very late in the game," Donoghue said. "We brought in huge amounts of money—$40 million—but this was a $120 million project."

But things didn't go smoothly. Big medical device manufacturers like Medtronic were only lukewarm on the idea and declined to invest in Andara. Then came the 2008 housing crisis, and when the FDA finally got back to Cyberkinetics, its regulators wanted more data. Andara would need another trial. "That was a major problem," Donoghue said. "If you're a company in that position at that time needing $3 million to do a trial that might lead to FDA approval? Virtually nobody in the world would give you money."

•

Cyberkinetics sold its rights to the Andara system in 2008. That same year, the company sold the rest of its assets to the Utah-based I2S Micro Implantable Systems Inc. for roughly $1 million. According to the terms of the sale, the Utah company could manufacture and sell the neural interface products, but Cyberkinetics retained the intellectual property and regulatory approvals. The company's stock was trading at a penny, and in October Donoghue and Nicholas Hatsopoulos resigned from the board. By the end of the year, the company's cash reserves were so low that it could only cover operating costs for another month. Cyberkinetics was going broke, and it still owed $600,000 to creditors.

Early investors such as Oxford Bioscience Partners lost their entire investment, as Cyberkinetics' CEO, Timothy Surgenor, worked to avoid bankruptcy. "We believed this was cutting-edge technology for the future," Jeff Barnes, a partner at Oxford, told *The Boston Globe*, adding that Cyberkinetics was one of the most difficult investments his firm had ever backed. "I still believe that the technology can be developed into a great product." Surgenor, who set up a consulting firm after Cyberkinetics collapsed, told the paper, "The ability to finance things that are exciting but have an unclear path to the market is just really tough. And BrainGate was the poster child for that."

Cyberkinetics sold the remains of the company to Jeff Stibel, a Brown graduate who founded Simpli.com, an early search engine he later sold for roughly $23 million. Stibel paid less than $1 million to acquire the BrainGate trademark. Included in the sale were more than thirty of the company's patent claims and the Cyberkinetics.com domain name.

Stibel said he planned to improve the system's software to support clinical researchers like Donoghue. But the Brown researchers had other plans. "He has no connection. We talked early on, and I told him I didn't want to be together. It just isn't the right thing to do," said Donoghue, adding that he and his colleagues had retained the essential intellectual property behind the technology, including Humphrey's BCI patent, Brown's decoding patent, and Normann's patent for the Utah array.

For Donoghue's colleagues, however, Cyberkinetics' failure was met with a complicated mix of relief, dismay, and no little schadenfreude. Nicolelis, who insists he has little taste for starting his own company, still appears to hold something of a grudge against Donoghue. For Nicolelis, Cyberkinetics' failure remains "a scandal."

For younger entrepreneurial researchers like Leuthardt, Cyberkinetics' failure was not just a cautionary tale. It was an obstacle that made investors skeptical of the field's ability to deliver. "Investors are dubious after Cyberkinetics. They say things like, 'Cyberkinetics didn't do so well, how are you different?' " he said. "There are a lot of dead bodies."

Still, there's no denying that Cyberkinetics cleared the way. It might have lost $40 million in the process, but the company won FDA approval to implant the Utah array in humans. "It's ridiculous, but they did it," said Schwartz. "They got devices made for human use. I give that to them. I give it to them today. They got it."

•

After the fall of Cyberkinetics, Donoghue and his colleagues transferred their research to Brown, where in a grand mansion of wood-paneled wainscoting and crystal chandeliers they reformulated the project as BrainGate2. Now in an academic setting, Leigh Hochberg, a cautious young neurologist with a clinical practice of his own, took on an expanded role. Hochberg still seems boyish in his midforties. His blazer is a size too large. His fingernails are raggedly trimmed, and he parts his fine brown hair to the side. He's the sort of man who rarely discards the formalities of academia, choosing instead to "limit our comments to the realm of things that are published." Seated in his upstairs office, he added,

"There can be times when you show your latest and greatest, and that's a good thing. But for me, the latest and greatest is something that appears in a peer-reviewed format."

Hochberg worked in Don Humphrey's lab as a graduate student in the 1990s. He had done his undergraduate work at Brown, but he shifted course when he interviewed with Humphrey, who was then listening in on the brains of monkeys while they played video games. During their first interview, the older scientist told Hochberg he wanted to see if they could use the neural information to re-create arm movements. "All of these questions came into my mind: Do we know enough about the brain to do this? Do we know enough about how the brain changes in cortical plasticity to do this? Was he joking?" Hochberg recalled. "I was interested enough to say that's exactly what I want to do."

Although he had known Donoghue at Brown, they didn't start working together until Hochberg became a neurologist at Massachusetts General Hospital, where Cyberkinetics brought him on as an independent academic investigator. "I never took a dime from the company," he said. "I was never a stockholder. I was never a consultant." Now that Donoghue had refashioned the project as an academic enterprise, however, Hochberg took an office upstairs, becoming the study's principal investigator.

From a scientific perspective, several fundamental hurdles still stood before the new study. Unlike Nagle, who received his implant a few years after his paralysis, Hutchinson had lived as a prisoner in her body for nearly a decade. Similarly, their second subject, Bob Veillette, did not join the trial until June 2011, some five years after his stroke.

It was unclear whether Hutchinson's motor cortex, after so many years of disuse, would have shed its duties as it was recruited for other brain functions. "Those were much more plausible outcomes than the fact that if you think about moving, it acts just like your arm was moving," said Donoghue. In fact, Hutchinson's motor cortex had remained intact. Her signals declined sharply after she was dropped, but that was due to the accident, not the implant, and for the first five years of the study, Donoghue pursued similar goals as he had with Nagle, working mainly with computer cursors.

By 2011, however, DARPA had chosen Schwartz to join its Revolutionizing Prosthetics program. With the program's vast cash reserves and goal to develop a brain-controlled prosthetic limb, Schwartz now had the resources to eclipse Donoghue's work with computer cursors.

The Brown researchers acted fast. In the coming months, they linked Hutchinson and Veillette to two separate robot arms: the so-called DLR arm, a sleek appendage on loan from the German Aerospace Center (DLR), and one of Dean Kamen's DARPA-funded DEKA arms. For five years, Hutchinson had used her BrainGate almost exclusively to navigate a computer desktop. Now the researchers linked her to the DLR arm for diagnostic testing, streamlining their decoding algorithm and testing the interface.

Six years after implantation, however, Hutchinson's electrodes had been through a lot. Not only was there the early accident, but also the hostile environment in the brain meant the signals that remained were not as strong. Despite these hurdles, Hutchinson managed to control the arms for four days, completing a series of reaching and grabbing exercises with foam balls.

Veillette, on the other hand, had received his implant less than six months earlier. His signals were much stronger than Hutchinson's, but he spent only one day linked to the DEKA arm performing similar reach and grasp exercises.

Ultimately, Hutchinson managed to touch her target nearly half the time with the DLR arm and nearly 70 percent of the time with the DEKA arm. Veillette was significantly more successful. In his forty-five attempts, he touched the foam ball about 95 percent of the time. Grabbing the balls was more difficult, and both Veillette and Hutchinson saw significant declines in their performance. Hutchinson managed to grab the ball only 21 percent of the time using the DLR arm and 46 percent of the time with the DEKA. Once again, Veillette was more successful: he managed to grab his target roughly 62 percent of the time.

The high point of Hutchinson's four-day session came when researchers placed a closed bottle of coffee on the work space in front of her. The idea was for her to pick up the bottle and sip from its straw.

Fully loaded, the DLR arm had seven degrees of freedom. For Hutchinson to pick up the bottle, however, researchers restricted the arm to the two-dimensional tabletop plane. They also restricted the hand to one degree of freedom, confining it to a simple open-close gesture.

The demonstration required Hutchinson to position the DLR arm before the bottle. Once the arm was in position, she initiated the grasp by thinking of closing her hand. This prompted the arm to execute a complex gesture—lowering the hand, grasping the bottle, and lifting it off the table. Once the arm had the bottle in hand, Hutchinson brought the arm toward her and positioned it by her mouth. She thought again of squeezing her hand, which this time prompted the arm to tilt at the wrist, tipping the bottle so she could sip from its straw. After drinking from the bottle, Hutchinson prompted the robot arm to straighten the bottle before she moved it back to the tabletop.

"This was the first time in nearly fifteen years that she had been able to pick up anything solely of her own volition," Hochberg said after publishing their results in *Nature*. "The smile on her face was something that I and I know our whole research team will never forget."

Once again, Donoghue and his colleagues had beaten Schwartz, becoming the first group to link an implanted human to a robot arm to perform a useful task such as eating.

•

By the time Donoghue and Hochberg arrived at the Neural Interfaces Conference in Salt Lake City, they'd already made the media rounds. Their research had been prominently featured in several national and international news outlets, and the pair held themselves aloof from the rest of the meeting, talking only to those who approached them while occasionally looking over their colleagues' posters or dropping in on a talk.

When it came time for them to present their findings, however, it was clear that the public's admiration had done little to blunt the sting of some of their colleagues' criticisms. A blog post about their research noted that the Brown team was one of "several groups involved in a highly competitive and sometimes vituperative competition to move this technology forward."

Hochberg was unhappy with the description and took his colleagues to task during his conference talk. "The things that some people are saying on background, it's coming across—it doesn't help us at all," he told the crowd, urging them to be more guarded when speaking to the media. "Objective criticism, of course, is the bedrock of what we all do. But as we are describing—complimenting, criticizing, whatever it may be—each other's work, a pause may be useful."

Then he took an oblique jab at Schwartz, whose team at Pitt had raised hackles a few months earlier when they sent out a press release about Tim Hemmes's work before publishing their findings. "Reviewed publications should precede press releases. Don't let your colleagues, your university press office, your company, or anyone else tell you otherwise," he admonished his colleagues. "I just encourage everybody just to adhere to what's really an ethical mandate, which is to do what we're supposed to do, which is to check our work with each other before we tell our favorite media representative about it."

Their colleagues had in fact been quite measured in their comments to the press about the BrainGate team's recent study. Nicolelis kept quiet, and Schwartz allowed only that the Brown group had shown the technology's "therapeutic potential," demonstrating "how a useful task could be carried out in a locked-in patient who had a long-term microelectrode implant." Privately, however, Schwartz was less impressed. "The first time I saw that was at a Nobel symposium in Stockholm. He showed that—and here's John Donoghue saying, 'We did it, self-feeding!'" Schwartz recalled. "I said, 'So, John, how many degrees of freedom is that control?' He said, 'Oh, about two and a quarter.' And then I get up and show seven degrees of freedom. I showed them the monkeys self-feeding. It was kind of lost on the audience."

Nicolelis was more scathing. "It's only five sessions. The woman had five years implanted. I have five hundred sessions in my monkeys, how can he have only five sessions?" he asked. He added that other technologies like EEG could deliver a similar level of control, initiating a movement in a preprogrammed robotic device. "She didn't need to be implanted to do that task, but for some mysterious reason nobody goes after that," he said.

Still, there was no denying it. Donoghue's team had pole position. Not only did they have their study site at Brown, but they were also enlarging their research team to include scientists at Stanford University, Cleveland's Case Western Reserve University, and the Cleveland FES Center.

Expanding their research across the country did not merely add to the team's brainpower. It also expanded their potential research subject population—a critical advantage in that it's rare to find participants who both qualify for the study and are willing to undergo brain surgery for a clinical trial. By expanding their research to Cleveland and Palo Alto, however, the Brown team had effectively tripled their potential research subject population.

What's more, by bringing the Cleveland team on board, Donoghue and Hochberg were taking a critical next step toward their ultimate goal. "The real dream is to reconnect the brain to limbs and to drive an FES device for somebody with paralysis," said Hochberg. "From our perspective, whether it's a robotic arm, a prosthetic limb, or an FES device, we want to use the flexibility and the power of the motor cortex to be able to drive those three things."

9. THE REDEEMER

Kneeling beside the second murder victim, Jan Scheuermann knew something was wrong. She'd written the script, coached the actors, and carried her props into the party, a charity fund-raiser and celebration of her tenth year organizing murder-mystery parties. There were some two hundred people in the room. She'd put the show on for free. She'd raised $1,000 for charity.

But as the mystery reached its denouement, she realized she was stuck. She'd felt weakness in her legs before, but she had always chalked it up to the cold. Now she knew it was something else. Her legs were unresponsive. She couldn't get them to move. It was February 1996, and Scheuermann, then living with her husband and two small children just north of Los Angeles, was quickly losing control of her body.

Her neurologist was stumped. He ruled out multiple sclerosis immediately, but he couldn't offer her a diagnosis. He sent her home. "I had no explanation for it," she said. "I thought it would go away."

Doctors at the University of Southern California tested her extensively for multiple sclerosis, performing a spinal tap and vision test. The results came back negative. "When something looks like MS and smells like MS, even though the tests are all negative, we classify it as MS," she recalled the doctors telling her. But like her primary neurologist, they could not offer her a treatment. They sent her home, too.

By summer, Scheuermann could no longer raise her arms to reach the bowls in her kitchen cupboard. Her body was failing fast, and at this rate she believed she'd be dead in a few years. She persuaded her husband to move the family back home to Pittsburgh. At least there, she reasoned, her children would be surrounded by family.

Like their colleagues in California, doctors in Pittsburgh ruled out MS, diagnosing her instead with spinocerebellar degeneration, a poorly understood set of symptoms caused by motor neuron death in the spinal cord. It's remarkably rare, an orphan disease. "It's not even popular enough to have an association," she said. "No one knows what to do."

Within three years, Scheuermann had gone from being healthy and active to a wheelchair-bound invalid. She could no longer walk. She had two young children to rear, but she was so weak she needed a motorized wheelchair just to get around. She couldn't feed herself, and she had to be placed on a shower chair for bathing. "I wasn't finding any light at the end of the tunnel here," she said.

Prozac blunted the suicidal thoughts that accompanied her decline, but now back in Pittsburgh, she was confined to her home. She had already given up her career and volunteer work. She couldn't help clean, and she could only attend parent meetings if the school held them in a building with ramps. Their Pittsburgh home had two stories, so Scheuermann slept on the ground floor, only rarely going upstairs to her kids' bedrooms. "I wish I could've taught them, this is how you fry hamburger, this is how you clean a pot," she said. "I missed tucking them in and watching them sleep."

Instead, a large portion of any given day for Scheuermann is consumed by the menial tasks of maintaining her inanimate body—having a caregiver bathe her, shampoo her hair, change her clothes, or help her go to the bathroom.

One of the idiosyncrasies of Scheuermann's condition was that while her motor nervous system was nonfunctioning, her sensory system remained intact. It wasn't just that she could feel everything. Some parts of her body were hypersensitive. It was painful, for instance, when the skin on her knee was stretched. Her body overheated easily, and it cramped if left too long in one position, often needing minor adjustments. "If I had the money, I would hire someone just to scratch me," she said. "Sometimes I can use the joystick I push to drive my chair to scratch my nose, but if no one's around and my head is itchy? It just itches."

Her condition taught Scheuermann to give caretakers precise requests. *Please move my right hand two inches up. Please remove the blanket from my right shoulder.* She tried to be what she called a "good invalid," grouping her requests so caregivers are not forever ferrying things back and forth to her, administering sips of water, turning on a fan, or adjusting an arm. "Every person that goes through this feels guilty for being a burden. People don't treat you like that, but you see the lengths they go to to get you ready to go, to strap you in, to get you to the van or whatever," she said. "You can't help noticing it."

Scheuermann's guilt wasn't confined to her paid caregivers. She was even more troubled by the emotional toll it took on her family. Her disease at times overshadowed her children's upbringing. It constrained her social life—a source of guilt for Scheuermann and a source of despair for her husband, Bob. "When she was healthy and we would come back to Pittsburgh, she had a lot of friends. When she came back sick and in a wheelchair, a lot of those friends disappeared," he said. "People will say, 'Oh, what's your wife doing?' I wouldn't tell them, but every now and then I would get close to someone. I'd mention my wife is quadriplegic, and you could see, sometimes they would take an emotional step back. How do you socialize as a couple?"

•

Scheuermann's faith in God had always assured her there was some grander purpose to her illness, but as the disease progressed, God's love felt increasingly distant. By 2011, twin bouts of septic shock had placed her in the hospital. The first nearly killed her. She was convinced the infection would return, increasing in severity. "That was the year I was going to die," she said.

Scheuermann was raised Catholic, and her entire childhood had unfolded within a three-block radius of St. Elizabeth of Hungary. But with its solemn masses and appalling sex scandals, the Catholic Church no longer provided her much comfort. She went to Mass, but only because that's where Bob went. And after two rounds of sepsis, her paralysis no longer seemed like part of some larger mysterious plan: it seemed like an awful genetic fluke. She was desperate, hopeless, and alone. "I told my family that the next time I got sick, I didn't want any antibiotics," she said. "I was just going to let the infection run its course."

Making matters worse, she was lonely after her younger child left for college. She had been working with a caregiver to adapt one of her murder-mystery parties into a novel, but she abandoned the project when her computer crashed, erasing her work. God felt very far away, and she spent hours each day watching DVDs, listening to audiobooks, or playing up to forty games of online Scrabble.

It was around this time that a friend sent her a video of a thin man in a wheelchair not unlike her own. Scheuermann watched as a young man used his brain to move a prosthetic limb, pulling pegs from a board and touching his girlfriend's hand. It was Tim Hemmes, of course, and it was unlike anything she'd ever seen.

Scheuermann was on a conference call with Schwartz's team at Pitt a few days later, kicking off a months-long procession of psychological and physiological tests. Researchers hooked Scheuermann up to an EEG machine to test her neural activations. They tested her verbal skills. They placed her in an fMRI machine and had psychologists evaluate her mental stability. "You know, fourteen years I haven't moved anything below the neck except for spasms—now I have this chance to move a robotic arm and to advance science?" she said. "What's the downside of that? There was nothing in me saying don't do this."

So it was that on a chilly morning in February 2012, surgeons implanted twin Utah arrays in Scheuermann's motor cortex—one above the area associated with hand movement, the other associated with arm movement. The arrays would only penetrate two millimeters into the brain, and surgeons counseled that the greatest risk lay in infection.

But Scheuermann wasn't worried. She'd spoken with Hemmes, who'd undergone a similar operation with no problem. How bad could it be? What she hadn't accounted for was the pain. "I woke up with a bad case of buyer's remorse," she said. "I'm thinking, Oh my God, I just had brain surgery for this! Am I nuts? Why didn't anyone stop me?"

As people, Tim Hemmes and Jan Scheuermann could not have been more different. Whereas Hemmes has the image of a pit bull tattooed on his neck, blue flames enveloping his forearms, and an eyebrow piercing, Scheuermann is given to wearing red librarian glasses and sweatshirts with plaid insignias. Just like Hemmes, however, Scheuermann proved a passionate research subject, meeting with Schwartz and his colleagues a few days after the surgery.

They were really just there to test the equipment, making sure the electrodes could pick up signals, but Scheuermann was ready to go. She arrived wearing a pair of mouse ears with a tail coming out from under her wheelchair's seat. "I was a lab rat," she said. Everyone in the room chuckled at the corny joke. Everyone, that is, but Schwartz. "I said, 'Come on, Andy. This is funny stuff. You should be laughing,'" Scheuermann recalled. "He said, 'Well, I don't think of you as a lab rat. You're a fellow researcher on this. I see you as a peer.'"

•

Schwartz had been working with monkeys for nearly thirty years, coaxing them to perform his neural bidding with a mixture of observation and rewards. But with Scheuermann, he had an articulate partner, one whom he could tell what he wanted. Better yet, she could tell him what she was thinking. There would be no controlling of cursors here. It was time to play the piano.

Scheuermann had always been a namer of things. When she felt guilty early in her sickness for holding people up, she named her legs Charlie and George. " 'Sorry for taking so long,' " she'd say, " 'but Charlie and George are not cooperating.' It just took all this off my shoulders." Now, as Schwartz plugged her into the APL arm for the first time, Scheuermann decided its name would be Hector, while she dubbed her twin electrode ports Lewis and Clark, "helping to chart the vast unknown areas of the brain."

Just as he'd done with his self-feeding monkeys, Schwartz first kept Hector under full computer control, recording Scheuermann's neural activity as she watched the arm move about the work space. As he recorded her neural activations, he slowly increased Scheuermann's control, attenuating parts of the signal that would send the arm in the wrong direction. They were giving her training wheels, shaving off errant signals while allowing her correct signals to guide the arm. As Scheuermann's brain adjusted to the algorithmic interface, the algorithm adjusted to her evolving neural patterns, creating a more efficient union between Scheuermann's brain and Schwartz's computers.

Schwartz never asked Scheuermann to use a classifier, thinking about making a fist to send the arm to the right or flexing her elbow to make it go left. He also didn't want her to imagine physical details like flexing her triceps to extend Hector forward or contracting her biceps to pull him back. He wanted her to move the arm naturally, and in those first days Schwartz surrounded the $70 million arm with pillows, almost like boxing bags. "You don't want to overthink anything, just punch the cushion, punch the cushion, punch the cushion. It makes it a lot easier for us to pick up the proper signals," he said. "If she's trying to come up with all of these complicated strategies and corrections, we can't understand it." The strategy let Schwartz know exactly what Scheuermann was trying to do. Her intention was transparent. She wasn't trying to break down the movement into confusing muscular actions or correct when Hector reached in the wrong direction. "I bet that she would be able to do 3-D movement on the first day," he said. "It took about an hour into the second day before she was able to do it."

Schwartz had never been able to work like this. He had always assumed his monkeys instinctually thought of reaching and grabbing as they controlled their robot arms, but he never knew for sure. And while his monkeys might have achieved levels of control well beyond that of other labs, his work always carried the disclaimer that it was confined to animal models. "Other people who have tried this in humans couldn't get anywhere near the kind of performance we got in monkeys. One of their excuses was, well, maybe humans can't do the same thing as monkeys," he said. "Not only can we do it in humans, but Jan learned incredibly fast."

Schwartz had hit his stride as a researcher, but he'd also lucked into an enthusiastic research subject. Schwartz was officially aiming for seven degrees of freedom. Seven would make the study a success. Privately, the researchers were aiming higher. "They said, 'Our personal goal is eight degrees of freedom,'" Scheuermann recalled. "I thought immediately in my mind, okay, my goal is ten."

Schwartz had been working for years before he finally managed to achieve stable three-dimensional wrist control in monkeys. Scheuermann figured it out in a matter of weeks. She went from 3-D control of the arm to 3-D control of the arm and wrist within the first month of the study. By the end of the second month, she was opening and closing Hector's hand. Sixty days in, and they'd already achieved seven degrees of freedom, the study's stated goal.

They had another ten months to go.

•

After Scheuermann's caregiver had warmed her hand pillows, placed a knit cap on her head, and wrapped her in soft fleece blankets, Scheuermann piloted her wheelchair into the cold winter morning to be loaded into the twenty-year-old van she and

Bob had modified—he with a wheelchair ramp, she with polka dots on the side. As the van's motor rumbled and its heat struggled to fill the cabin, Scheuermann lay strapped in her reclined chair, staring at the ceiling as the van rolled down the steep roads of Whitehall Borough toward the University of Pittsburgh Medical Center.

Her caregiver, Karina Palko, was driving them into the city, where Scheuermann was scheduled to test an orthotic for possible use in the study and later to talk to a spinal cord injury support group. Nestled in a pile of manila folders on the front passenger seat was a stack of bookmarks for *Sharp as a Cucumber*, the murder mystery Scheuermann, with Palko's help, had resurrected and published on Amazon.com. She planned to talk to the group about the Schwartz study. Maybe she'd give a short plug for her book, but mainly she wanted to emphasize how important it is that people with spinal cord injuries be involved in research—not just for the advancement of science, but also for their own sense of purpose and self-worth.

Growing up Catholic, Scheuermann was committed to the ideal of service. Hunger had been her main cause, and throughout her life Scheuermann had volunteered in canned-food drives and soup kitchens. She had always clipped coupons for deals on things like deodorant and soap, and in California she'd come up with a coupon scheme to help the needy. "I had about twenty unopened deodorants. I thought, well, I don't need all of these, so I put them in a bag and took them to St. Vincent de Paul," she said. "Then I started thinking, if I had more coupons, I could do so much more."

She called her nascent organization the Redeemers, asking other parishioners to bring in their unused coupons. Scheuermann would then sift through the coupons, looking for the best deals. In the mornings, she'd stop at a store while taking her kids to school. "I would give them each three coupons. So the three of us would show up at the checkout lane, each of us with three cans of vegetables and three boxes of cereals and the coupons." After taking the kids to school, she'd stop by the store again, redeeming three more coupons. She repeated the drill each afternoon when she picked up the kids. "My kids were accomplished coupon shoppers at four years old," she said. "I was getting canned vegetables and a lot of other things besides deodorant by the time it was all over."

Then came the weakness in her legs and the move back to Pittsburgh.

"My mom pulled out a big manila envelope full of receipts from my work at St. Vincent de Paul. I just burst out crying," she said. "That's what I missed. I loved being able to do that, make a difference."

10. BLIND SPOTS

Scheuermann was back at the Schwartz lab the morning after her meeting with the support group. An assistant sterilized Lewis and Clark with Q-tips and hydrogen peroxide before plugging Scheuermann into the computer. They were months into the trial by then, and Scheuermann's brown hair was tinged blond from all the peroxide. Dressed in a sweatshirt and covered in a polka-dot fleece blanket, she wore a pair of dark 3-D glasses as the neuroscientists calibrated her brain with the computer. Hector stood immobile to her right as Scheuermann watched a monitor mounted on the wall in front of her. The screen showed a virtual Hector, silver with a bluish hue, against a black-and-white checkerboard background. A red oval appeared in one section of the screen. As a bell rang, the synthesized voice of a woman instructed, "Wrist up, spread."

"Uh-oh," one of the graduate students said.

"Grasp," the synthetic voice commanded.

The virtual Hector was under computer control. Everyone in the room expected the arm to move swiftly toward the red orb, but the simulated arm just sat there, frozen like its physical counterpart.

"Wrist clockwise," the calibration module continued. "Release."

Again, nothing.

"It's not running," one of the graduate students said as he peered at a graph of action potentials and tapped on the keyboard. "It would almost have to be the executive, wouldn't it?"

"Reset the executive, yes," said Brian Wodlinger, a postdoc charged with maintaining the algorithm.

"All right, we're going to try this again."

"Wrist counterclockwise, thumb in," the module droned. Again, the virtual Hector just sat there.

"I don't understand it," said Wodlinger.

"I don't know what's happening," seconded the graduate student.

"Grasp," said the calibration module.

"Stop," said Wodlinger.

"Wrist clockwise," said the module.

"Something's got to be running," he said.

"Oh!" the grad student barked, slapping a palm to his forehead. "I opened the wrong environment. Here we go." Hunched on their task chairs, the researchers studied the screen, opening new windows and pressing keys. Everyone else in the room waited. The virtual arm stood still.

"Okay, sorry for the technical difficulties," Wodlinger said.

"It's fuzzy again," said Scheuermann.

"Here we go," said the graduate student.

"Wrist counterclockwise, spread," the woman's voice commanded. Now the virtual Hector moved crisply toward the red oval. "Grasp," it commanded when it reached the target. "Wrist clockwise," the module instructed as a green rectangle appeared in another area of the screen. The arm moved swiftly toward the rectangle, traveling across the screen toward the new target. "Release," the voice commanded. The arm disappeared, reemerging a few moments later at its starting position.

The virtual arm completed this exercise seventeen times. Meanwhile, Scheuermann sat immobile in her chair, staring up at the screen as researchers recorded from her motor cortex. Just as they'd done in the initial training sessions, the scientists correlated Scheuermann's neural activity with the movements of the virtual arm, linking her neural response to specific gestures before giving Scheuermann control. Her neurons were adjusting to the computer interface. The algorithm was adjusting to her.

Once they had their initial decoder, the bell chimed, and the female voice led Scheuermann through a series of more delicate hand movements: moving the wrist, pinching, and spreading the fingers. The computer still had some control, but Scheuermann was taking over. The system's training wheels still suppressed input from errant neurons, but they allowed neurons that fired appropriately to drive the arm's movement. The virtual arm had moved steadily under computer control, but Scheuermann's movements were more hesitant, as researchers refined the decoder.

The scientists still weren't satisfied, and they hunched over their computers to put the finishing touches on the decoder. Scheuermann, meanwhile, asked Palko to bring her *The New York Times* Sunday crossword puzzle. Dressed in a Winnie-the-Pooh and Tigger hospital smock, Palko accompanied Scheuermann to each of her testing sessions. She was forever alternating between making minor adjustments for Scheuermann—raising her chair back a few degrees, moving a pillow—and reading a romance novel she'd checked out from the library. Palko took careful notes on the study's progress, and she was quick to anticipate Scheuermann's needs, producing the crossword puzzle on demand.

As Scheuermann puzzled over 69-Down, Schwartz and Wodlinger huddled together over a puzzle of their own: a drift had apparently crept into the arm's movement.

"It's a difficult problem because the mean is below half height so that it should be negative, and firing rates can't be negative," said Schwartz, dressed in khakis and an open-collared sweater. "So you truncate it, and that's what causes the drift probably, so you need the bias. One way, a crappy way, to correct for it is the bias that we've been doing. But there's a better way to do it, and that's using some nonlinear functions."

"But if it's nonlinear, then you can't balance in the center," said Wodlinger, who apparently understood Schwartz's diagnosis. "That's the problem."

"That is the problem," said Schwartz. "What that means is that when you try to use the population vector, it won't work very well. You should talk to Steve. He's working on it."

"You can't just get the new algorithm," said Wodlinger.

"Well, talk to Steve. There are two problems: one is that, and the other is the limited sample size."

A few feet away, Scheuermann was equally stumped as she peered under her 3-D glasses at the Sunday crossword. "One more question, guys. 'Gold-compound salt'—is that auxite? *A,U,* what? I know the *A,U,* blank, blank."

But the researchers were ready to give her full control, and the virtual arm now charted a shaky course toward its target, hovering noncommittally before finally hitting the mark. It had taken nearly an hour for Schwartz and the other PhDs in the room to build their decoder. But as tedious and error prone as these exercises were, they were also absolutely necessary to get this level of control.

It's one of the major challenges facing neuroprosthetics. Although the Utah array is designed to float with the brain, it simply doesn't shift as readily as the soft, custard-like organ it penetrates. An electrode that one day is snuggled against a neuron will the next be farther away, making action potentials harder to record. A new neuron might have migrated toward the electrode, and that neuron will have different synaptic connections and firing patterns. "You end up recording from different neurons every day," said Jennifer Collinger, one of the study's coordinators. "Probably one-half to two-thirds stay the same from one day to the next, but the others are different, so it's not like she has a stable population of neurons."

Then there's the variable nature of neurons themselves. Each cell in the motor cortex has a preferred firing direction, but these preferences are fluid. They change from day to day, even from hour to hour. A cell that fired at 80 hertz on Monday will fire at

60 hertz for the same task on Tuesday. The old algorithm is no good and needs to be frequently updated to keep pace with the brain's shifting landscape.

This was particularly true in the early days of the study, when Scheuermann's neural population was so unstable researchers sometimes had to recalibrate after only an hour. "We just had to start over," Collinger said. "Things would change so dramatically."

The implant eventually stabilized, but they still had to calibrate the system daily—no mean feat considering that it takes a bevy of computational wizards about half an hour to build the decoder. It's a big hurdle facing any commercial neuroprosthetic that uses penetrating electrodes. "If you could come up with a five-minute calibration, people would probably be willing to do that," Collinger said. "But if you had to go through this whole twenty- to thirty-minute process that is computationally intensive? That's not going to be a viable device."

•

But calibration raised other, more mysterious questions as well.

When they began the study, researchers used Hector to build their decoder. It was tedious but effective, as long as Scheuermann had only three or four degrees of control. As they increased Scheuermann's control, however, something remarkable happened: though she could easily open and close Hector's hand in free space, the arm would freeze when she tried to grab a plastic cone. "Jan, what the hell. You just did it," Schwartz would say. "Then we'd say, 'Jan, close your eyes.' Boom! She could close her hand around the cone. As soon as she opened her eyes, she couldn't do it anymore. What the hell is going on?"

It seems pretty straightforward, after all: we grab a thing by extending our arm and adjusting our grip to suit the object. As Hector can attest, we understand those mechanics fairly well.

The real question is what happens in the brain. We expect our senses to give us an accurate portrayal of the physical world. We expect that when we open our eyes, we behold the physical world as it actually exists. We learn to expect physical objects to behave in a consistent manner. If we tilt a cup, we expect liquid to pour from it. We expect light to shine from above and a ball to drop at a certain rate. In other words, we expect the world to adhere to specific physical laws.

Evolution has hardwired our brains with an intuitive capacity to understand those laws, and that ability, augmented by ongoing experience, enables our brains to make all sorts of predictions about how objects around us will behave. We're able to cross a street safely because our evolutionary inheritance has endowed us with powerful models to accurately predict how long it takes objects to reach us.

But biological evolution only goes so far. It never prepared us for large land-based objects that can travel at constant speeds of seventy miles per hour. For that, our inherited internal models must be modified by a lifetime of watching cars whiz by on the interstate. It's this two-pronged approach that allows us to be successful in the world, whether it's crossing the street or learning to lead a pheasant with a shotgun.

But is it really so simple?

At the dawn of the nineteenth century, the Scottish neurologist Charles Bell hinted at the notion that the brain does not directly perceive physical reality. Bell is best known for discovering the nervous system's structure of discrete motor and sensory nerves, arguing that specific nerves are sensitive only to specific stimuli—you can't hear with your fingers, for instance. But Bell took the idea one step further, reasoning that "neither bodies nor the images of bodies enter the brain. It is indeed impossible to believe that colour can be conveyed along a nerve, or that a vibration . . . can be retained in the brain; but we can conceive and have reason to believe, that an impression is made upon the outward senses when we see, hear or taste." Our experience of the outside world is not direct, Bell argued. Rather, the outside world bombards us with stimuli, but what we experience is a neural representation of that world.

The German physiologist Johannes Peter Müller furthered Bell's line of reasoning when he argued that nerves have "specific irritability" and "specific energies." Müller was grappling with the idea that we can perceive similar sensations from wildly different stimuli: the eye can perceive light, electrical current, and even physical pressure as "vision." "The immediate objects of our senses are merely particular states induced in the nerves, and felt as sensations," Müller wrote in the 1840 edition of his seminal text, *Handbuch der Physiologie.* It isn't merely that the brain relies on neural translations to accurately perceive the outside world. Rather, the outside world presents fraudulent stimuli, tricking the peripheral nervous system and causing the brain to perceive a false reality. It doesn't matter whether the neural signals are instigated by pressure, light, or electricity: the brain—we—*experience* these stimuli as "light." The world is not necessarily as it appears. We can't always trust our senses. Our brains can be fooled.

This revelation remained something of a parlor trick through much of the twentieth century as brain scientists mined the notion that sensory perception worked like an assembly line. Namely, the peripheral nervous system streamed raw data into the brain, which refined the information, pulsing it through a series of higher brain processes until it produced a conscious perception of reality.

Take vision. The prevailing twentieth-century theory was that as light signals bounce off an object, they bombard the retina, a sheet of specialized photoreceptors at the back of the eye. Those specialized retinal cells then go to work on the photons, with specific classes of retinal neurons sensing colors, movements, shapes, lines, shadows, and so on. These retinal neurons translate the information into action potentials that stream along the optic nerve en route to the thalamus, a deep-brain structure that acts as a sort of sensory switchboard, routing sensory information to the appropriate cortical regions.

In the case of vision, the thalamus directs ocular information to the primary visual cortex. As the visual stream jumps from neuron to neuron, the higher cortical regions synthesize these smaller bits of information (light, shading, lines, and so on) into specific features (the precise shape of the eyes, a specific color of the skin), ultimately perceiving them as an immediate conscious reality.

But for all the prodigious neural power this model attributed to the cortex, it also viewed the brain as essentially responsive. Its neural architecture at the ready, this complicated neural assembly line would spring to action only when it received input from the peripheral nervous system.

No doubt, the model made sense. After all, not only does the physical world appear to us as objective, but our perception of that world seems instantaneous, seamless, and often unequivocal. We see a man walking a golden retriever along a tree-lined street, and we immediately behold the reality of the scene. We bite into a piece of chocolate cake, and we are bombarded by its soft feel and sweet taste. The world and all its stimuli seem to inundate us, their obstinate physicality unmistakable. It's an impression that is only reinforced by our motor system, which seems no less fluid. We reach to grab a peach, and we don't think of the chorus of muscles we must first engage. Our hand simply wraps around the fruit, often before we are consciously aware that we've even reached. We are reacting to the world around us, seeing and tasting and feeling objects that appear before us.

As intuitive as it sounds, neuroscientists such as David Eagleman and V. S. Ramachandran have recently argued that this stimulus-response theory fails to explain even some of our most basic lived experiences. As Eagleman argues, its veracity was challenged in the 1950s, when a neuroscientist named Donald MacKay hypothesized that the primary visual cortex builds internal models to predict information arriving from the optic nerve.

Rather than relying on a one-way tract of communication, the brain enlists feedback loops, where higher brain regions exchange information with lower regions. The theory was dramatically bolstered when MacKay found ten times as many input fibers running from the primary visual cortex to the thalamus as there are fibers running from the thalamus to the visual cortex.

We've all seen line drawings of a cube like the one below. We stare at it for a moment, and we notice how our perception of the image begins to shift. One moment, the "front" side of the cube appears oriented down and to the left; the next, it seems pointed up and to the right. Nothing has physically shifted, and yet we find the image ambiguous. Our eyes, or rather our brain, can't make sense of the visual stimuli, so our perception shifts back and forth. "There's a striking point here," writes Eagleman in his book *Incognito*. "Nothing has changed on the page, so the change has to be taking place in your brain."

If the brain were merely taking in objective stimuli and refining them into conscious experience, we wouldn't be so confused by these twelve straight lines. The object would present itself, and the brain would perceive it. But the brain actively constructs vision, using its formidable neural architecture to process incoming stimuli and build consistent models to produce a coherent experience of the objective world. In the example below, simple as it is, the visual stimulus is ambiguous. There are two warring interpretations, and so our brains, unable to decide which is definitive, toggle between the two, perceiving the image downward facing one moment and upward facing the next.

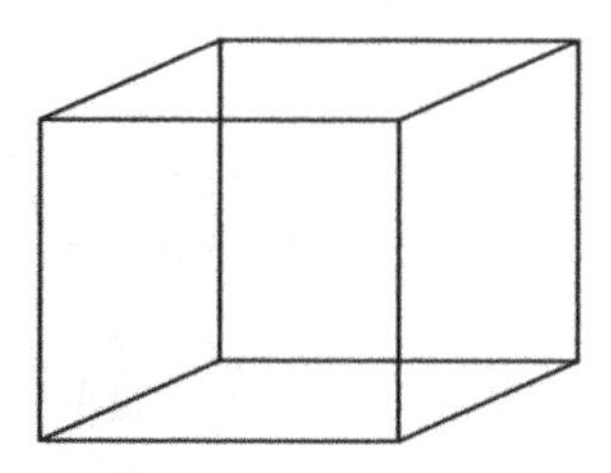

But as Eagleman notes, perhaps the best example of the brain actively constructing vision is the blind spot we have at the back of our eyes. It's the sort of thing you could go a lifetime and never notice. And indeed, plenty of us did until the seventeenth century, when the French physicist Edme Mariotte discovered we have a large blind spot in both of our eyes. If the brain simply processed incoming visual information, then it would stand to reason that we would notice these blind spots. There would be a large void in our visual field where no incoming information was available.

But we don't perceive a hole in our vision. Part of the reason for this is that the blind spot, located where the optic nerve attaches to the retina, is in a different part of each eye, enabling us to cover the visual field with overlapping stereoscopic vision. Even so, close one eye, and no gap appears.

Neuroscientists contend that we don't notice it, because the brain actually fabricates "vision" over the area of the blind spot, constructing a pattern or model that mimics the prevailing background. It's not that the brain simply responds to visual stimuli. Rather, the brain interprets incoming visual data as it actively builds vision.

Our motor system works along similar principles. It may seem that our experience of the outside world is instantaneous, but it actually takes milliseconds for external stimuli to transform into action potentials that move neuron by neuron from the peripheral nervous system to the brain, where we experience them as touch or sight or taste. That lag, infinitesimal though it may be, would make it impossible for us to successfully navigate the world if we were simply responding to stimuli. Take swinging a racket to hit a tennis ball. As Eagleman notes, if we were merely reacting to stimuli, we would always be a few hundred milliseconds behind the ball. Of course, we are able to hit tennis balls for the same reason we are able to cross the street safely or feed ourselves. Not only are our brains hardwired to learn the laws of physics, but that understanding is further refined through a lifetime of watching objects arc through the air with gravitational pull. Together, they enable us to make predictions about when and where the ball will arrive. "When the brain creates a model, it anticipates what it is going to view in each moment in time. Before you touch something, or before you open your eyes to see something, the brain is already creating an image of what you are about to see," said Nicolelis, adding that there is a critical difference between a brain in a sleep state and one that is awake. "For a long time, neuroscientists only recorded the brains of anesthetized animals. For the last ten years, we are starting to recognize that state dependence is a key issue in understanding how the brain works." We don't simply respond to sensory stimulus; we rely on internal prediction models. Or, as the hockey great Wayne Gretzky once quipped, "A good hockey player plays where the puck is. A great hockey player plays where the puck is going to be."

In that sense, we are all great hockey players when it comes to the physical world. Of course, these models are not fixed. They are constantly updated by a continuous stream of sensory information, which enables us to adjust our models to more accurately predict what is about to happen.

But these higher brain regions don't simply build models to compare with input streams from the peripheral nervous system. To the contrary, neuroscientists believe the brain's give-and-take is so robust that these higher brain regions can actually affect the lower brain regions, priming them to experience external stimuli in a specific way. Each time you imagine the taste of melting butter on freshly baked bread, those lower areas of the brain engage as if you were actually experiencing the situation. Similarly, take the image below. It's not an optical illusion. It's actually a poorly developed old photograph. So what is it?

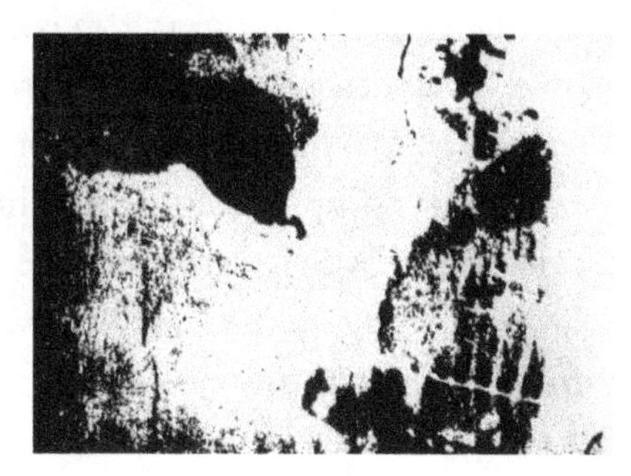

Do you see something resembling a grenade in the lower right-hand corner? Perhaps the beginnings of a pair of faces in the upper right-hand corner? It's tough to say. Though we try to make sense of the image, the brain has little expectation for what it should be seeing. It can't build a model, and so it cannot interpret or adjust the stream of visual information pouring in through the retina. We try to discern a coherent image, but most of us will be left seeing nothing but an abstract pattern in black and white. Once we hear the words "calf's head," however, we have a context. With that little hint, our brains quickly build a model, and the image snaps into focus.

Our brain, whether it be the sensory system or the motor system, actively constructs the world as it expects it to exist, orientating our bodies in the physical world. What emerges from these multiple complex systems is an all but seamless conscious experience, one that only begins to break down when confronted with ambiguous stimuli.

•

It was perhaps just this sort of uncertainty that appeared when Scheuermann found herself incapable of grasping the cone. Closing Hector's hand hadn't been a problem when she controlled the arm with fewer degrees of freedom. But when researchers increased her control complexity, the hand began to freeze each time she tried to grab the cone. More mysterious yet, the problem disappeared when Scheuermann closed her eyes. Without visual input, she could easily grab the cone, but the moment she opened her eyes again, Hector would freeze in place.

This was no engineering problem. Hector was fine. He moved gracefully under full computer control. The problem here was neural, and the question for Schwartz became whether learning to interact with objects engaged a different set of brain processes than simply making an arm move.

If all Schwartz had wanted to do was make Hector move, he could build a simple neural decoder like the one Wang had used with Hemmes. Such a decoder would enable Scheuermann to consistently make certain actions, but it would also be limited. Scheuermann would only be able to perform a set number of programmed movements. She wouldn't be able to tailor her actions to specific circumstances or objects. Schwartz wanted something more refined. "You don't just have a few cells that turn on to move to the right," he said. "The whole population turns on. Some cells fire less; some cells fire more. But it's that combined activity that is important and generating the movement." Schwartz wanted to craft a brain-machine interface capable of extemporaneous movement so Scheuermann could explore physical space intuitively, reacting to specific objects with unique grasping patterns.

His earlier work with Georgopoulos had shown the importance of neural populations and preferred firing directions. That was fine as long as the control complexity wasn't too advanced, but this new neural hiccup had him thinking something else was at play. "There was something specific to the object that was buried in the neural signal that we were missing," he said. "Instead of thinking about the way the hand is represented, maybe we should be thinking about what the object is that's represented and what's going to be done with the object."

The brain may build lush models of the physical world, replete with expectations of sensations and how our actions will affect objects. But it's one thing to observe the neural storm that occurs as the brain creates a model. It's quite another to try to understand the thousands of spiking neurons within that storm and translate them into a meaningful brain-computer interface.

But that was the challenge Schwartz saw before him.

In his work with hand shaping in monkeys, Schwartz had attached tracking sensors to the animals' hands and arms. The idea was to record how the limb physically moved as it prepared to interact with specific objects while he also recorded the associated neural activity. Using information from limb movements alone, Schwartz found that he could predict with 99 percent accuracy what object the monkey was reaching for.

But it was in the brain where things got interesting. "I can already do object recognition based on motor cortical activity. Not only that, but I can tell you how different the brain thinks different objects are. So you may think that a cylinder and a sphere are different objects, but I can look at the neural activity and say, 'Well, the brain doesn't think those are very different,'" he said. "But that's right at the edge of what we know how to do. We're exceeding the bounds of our scientific knowledge."

Just how the brain recognizes distinct objects remains a mystery, but a decade earlier the UCLA neurosurgeon Itzhak Fried uncovered a tantalizing clue while performing brain surgery. With the patient under anesthesia, Fried removed a portion of the skull, exposed the brain, and implanted electrodes in the medial temporal lobe, a brain region essential for the conscious memory of facts and events. Once the electrodes were in place, Fried woke his patient, a fairly common surgical technique that enables surgeons to map the brain during surgery. Before proceeding with the operation, however, Fried showed his patient a series of photographs.

The researchers wanted to see how the brain reacted to pictures of people it recognized versus images of people it had never encountered. Accordingly, some of the photographs were of well-known celebrities; others were of complete strangers. Fried monitored many neurons as he scrolled through the images, but he noticed that a particular cell spiked like mad whenever the actress Jennifer Aniston appeared on the screen. The neuron quieted down as soon as they removed the photograph. The cell didn't spark to attention at the sight of equally well-known celebrities like Julia Roberts, Bill Clinton, or Halle Berry. Similarly, it was equally unimpressed by photographs of strangers. Intriguingly, the neuron didn't fire when the monitor displayed images of Aniston with Brad Pitt, her husband at the time. But then Fried would present a picture of Aniston again, and the neuron would spike to life.

The brain has hundreds of thousands, if not millions, of neurons spiking at any given moment. But in this instance, the cell spiked only in the presence of Jennifer Aniston, leading neuroscientists to wonder if the neuron was somehow tuned exclusively to the actress. If so, it stood to reason that there would be similarly tuned neurons for everyone a person knew or recognized. And indeed, researchers went on to find neurons that spiked only in the presence of images of Halle Berry, Kobe Bryant, and Julia Roberts. Still, the question remained: Did the brain devote specific physical spaces—individual cells—to the recognition of particular people, or was something else at play?

The so-called grandmother neuron (shorthand for the idea that individual cells are responsible for recognizing individual people) is a problematic concept. What would happen if your grandmother neuron died? Would you simply be unable to recognize your grandmother?

Perhaps more important, though, neuroscientists such as MIT's Sebastian Seung have argued that our general ability to perceive and recognize individual faces is far too complicated a task for a single neuron to perform. They hold instead that cells like the Jennifer Aniston neuron are embedded in neural networks, a piece of a puzzle that comes into place once the others are assembled.

At any given moment, an individual brain cell receives chemical signals from hundreds if not thousands of connected neurons in its network. As excitatory and inhibitory neurotransmitters bathe the receiving cell's receptors, they coax the neuron to fire or to remain inactive. Essentially, the receiving neuron reaches a tipping point, a magic combination of inputs that prompts the cell to fire. A pulse of electricity courses along the cell body, which releases neurotransmitters of its own, encouraging nearby neurons to fire or remain at rest.

Neurons don't fire in isolation. They fire in networks, and scientists like Seung believe it is the populations' ephemeral spiking patterns (and not the firing of any one particular cell) that underlie perception and memory. A network of several million neurons could create a near-infinite number of firing patterns, enabling the brain to recognize countless people, and that's to say nothing of different fruits, dogs, leaf patterns, and so on. A neuron's spiking, then, is wholly dependent on information it receives from the network's momentary configuration.

After all, we each exist and think of others in a plethora of wildly different contexts. There's the physical, which might be someone's hair, the color of his eyes, or the shape of his nose. Add to that the sound of a person's voice, his gait, and his distinct movement patterns. But there are also less tangible aspects we associate with a person. For a celebrity like Aniston, those might be her failed marriage to Brad Pitt or her role as Rachel on *Friends*. "Let's describe Jen as a combination of simpler parts," writes Seung in his book *Connectome*. "She has blue eyes, blond hair, an angular chin, and so on (as of this writing, anyway). If the list is long enough, it will uniquely describe Jen and no other celebrity." In Seung's telling, the Aniston neuron is the final link in a neural network. The cell's firing threshold is extremely high, and it must receive synapses from all of the contributing neurons—the right blue-eye neurons, the blond-hair neurons, and the angular-chin neurons—lower down the chain. "It spikes only when all of the part neurons spike, a unanimous vote that happens only in response to Jen. In short, a neuron detects Jen as a combination of Jen parts, which are detected by other neurons." Seung's theory of perception is hierarchical: the blue-eye neuron receives synaptic inputs from neurons tuned to ever simpler parts—a neuron for the white of the eye, a neuron for the pupil, the iris, the eyelid, the eyelash, and so on—eventually reaching neurons that are tuned to stimuli that can't be further reduced, like horizontal lines and spots of light and shadow. These simpler neurons are not devoted exclusively to Aniston. To the contrary, they're at the ready to register blues in the sky or the blue irises of others. It's only when a magic combination of these lower neurons fire that we register Aniston. The so-called Aniston neuron fires, then, not as the sole witness to the celebrity but more like a light that flashes vigorously in response to a specific pattern of lesser lights.

Be that as it may, researchers have found that cells like the Aniston neuron do not fire only when a person sees a photograph of the person in question. In subsequent experiments, Fried found these neurons fired with equal intensity when he simply displayed the name of the person on the computer, suggesting, perhaps, that the cell makes no distinction between the symbolic representation of the person and the sight of the person.

•

Fried and Seung are talking about perception, recognition, and memory, but for Schwartz the Aniston neuron might point to a neural principle that applied equally to movement—not only for the immediate question of why Scheuermann was unable to grasp objects, but more fundamentally how we conceive and interact with the physical world. "The question is whether Jennifer Aniston is built up of features—a certain shaped nose and distance between the eyes—and it's this conjunction of features that leads to Jennifer Aniston. Or is it a symbolic thing?" he wondered. "Do we have a symbolic idea of Jennifer Aniston?"

Seen slightly differently, this was the same question he was asking about objects and how we interact with them. Simpler brain-computer interfaces concentrated on the limb, assigning various movements to specific neural patterns. But those interfaces could only perform a limited number of movements. Schwartz had achieved unprecedented control over robot arms using population vectors. But as the arm's control complexity increased, he was no longer certain his tried-and-true method could accommodate the brain's symbolic conception of the physical world.

He was becoming convinced that he had to think of movement in increasingly abstract, or symbolic, terms. Instead of merely looking at the preferred firing directions and populations of neurons, Schwartz wondered if he needed to take intention into account. Namely, how did Scheuermann plan to interact with an object? After all, we'll grab a box very differently if we plan to open it as opposed to lift it, and the moment we begin reaching for a box, we are already planning our trajectory. Not only are we orientating it in relation to ourselves, but we are also determining how much force we will use, the speed of our movement, and how to shape our hand. We're simultaneously forming expectations of how the object will respond: Will the bottom of the box sag? Will it be warm? Will it be firm? Will opening the top require all of our fingers? Will we need one hand or two? "It would be like what your intended use of the object is," he said. "If I was going to pick up a cup to drink, I might pick it up more gently. But if I'm going to throw it, I'll pick it up differently. That sort of symbolic action is probably represented in the brain." The question, then, was not merely whether the brain creates specific representations for different objects. Rather, does the brain create different representations *for the same object*?

Just as the brain makes predictions to fill blind spots in our vision, so too does it create predictive models as we physically interact with the outside world. Those models are never objective; they depend entirely on how we intend to interact with the physical world. This experiential preview is continuously updated by sensory information, but as long as the sensory information doesn't contradict our internal model, we rarely become aware of the complex coordination of muscles and joints and bones it takes to interact with the physical environment.

Think of climbing a flight of steps. We rarely think about how high to lift our leg, how much force to exert on our foot, or the incline of our body against the stairs. We've spent a lifetime climbing steps, and we're on autopilot as we depend on our brain's symbolic representation of the physical environment to carry us through. But occasionally we miscalculate. We think there's another step at the top or that the rise is slightly lower. We stumble, and as the sensory information streaming up our peripheral nervous system contradicts our internal model, conscious awareness comes online.

"That's the job of sensation," said Schwartz. "What you're doing is always predicting what's going to happen in the future."

But as long as the physical world conforms to our expectations, we're happy to go about our business—driving to work with no memory of how we got there, absentmindedly taking a sip from our water bottle—with little conscious awareness of what our bodies are doing.

Could it be that Scheuermann, who could watch but received no sensory feedback from Hector, was unable to grasp objects because her neural model was being contradicted by a lack of sensory feedback? After all, with her eyes open, she could see the arm move, but this visual information wasn't accompanied by sensation. Did that somehow interrupt the internal model? Faced with the bare physicality of the cones, perhaps Scheuermann was simply thinking too consciously about the grasp. Perhaps her conscious brain was confusing the computer by overriding her internal neural model.

There was no question that motor neurons are associated with movement. But it was becoming increasingly clear to Schwartz that the motor cortex and our physical interaction with the world were inextricably linked to intention and the creation of predictive models. "We have these ideas that neurons fire in a certain way because they're trying to move an arm, but that's an assumption on our part," he said. "What if that's not really what they're doing?"

11. FROZEN MIRRORS

Much has been made of mirror neurons since Giacomo Rizzolatti first discovered the cells in his Parma laboratory. Rizzolatti was recording from the brains of monkeys, and researchers have since found similar cells in several areas of the primate brain, including both the sensory and the motor cortices. What's remarkable about mirror neurons, and what they're named for, is their ability to re-create what we see in the world. When we wince at the sight of a vicious tackle, mirror neurons in our sensory cortex come to life. When we're at the edge of our seats during a tennis match, mirror neurons in our motor cortex are firing away. These neurons, which some scientists believe account for 20 percent of cells in some areas of the macaque motor cortex, do not merely allow us to experience vicariously the crush of a tackle or the smash of a ball.

Rather, as part of the motor cortex, they are among the population of cells that spike each time we ourselves hit a tennis ball. They are not mere witnesses. They are members of the neural population associated with the actions themselves. And scientists theorize that by spiking at the sight of a tennis player hitting a forehand, these cells at some level re-create the swing itself, enabling us to actually experience the swing *as though it were our own.*

We may not have Rafael Nadal's refined motor skills to expertly hit a forehand, but we are all familiar with the gross physical action of swinging. We make those sorts of movements throughout our lives. We can approximate Nadal's movement, enabling us to understand and identify with his actions. In that sense, to watch Nadal play tennis is (at least neurally) to play the game ourselves.

But neuroscientists believe the cells' re-creation of physical movements and sensations is not confined to large physical gestures. These exquisitely tuned cells are also sensitive to the subtlest shift in another person's face and demeanor. When we see someone's eyes tighten in distress, or the corner of his mouth turn up in flirtation, we are right there with him, "mirroring" his emotion. We understand his intentions and state of mind because, as Bill Clinton would have it, we feel his pain.

This facet of the mirror system has prompted many neuroscientists to theorize that it is the cellular foundation of human empathy, an essential building block of our shared humanity. "Mirror neurons undoubtedly provide, for the first time in history, a plausible neurophysiological explanation for complex forms of social cognition and interaction," writes the neuroscientist Marco Iacoboni in his book on the cells, *Mirroring People*. "By helping us recognize the actions of other people, mirror neurons also help us to recognize and understand the deepest motives behind those actions, the intentions of other individuals." In essence, these cells offer a biological answer to a question that has vexed philosophers for centuries. Namely, how do we know that other people have beliefs, emotions, knowledge, or desires that are distinct from our own? And while we're only just beginning to understand the role these cells play in our understanding of other people, Iacoboni is not alone when he argues that the mirror system is the biological mechanism behind "simulation theory," the philosophical concept that we recognize and understand the mental states of others (their intentions, their desires, and their beliefs) by imagining them ourselves.

Of course, it's one thing to recoil when we see footage of the University of Louisville basketball player Kevin Ware collapse in agony as his tibia snaps in half. It's quite another to develop a nuanced understanding of another person's mental state. So far, the bulk of the research on mirror neurons has been performed in monkeys, where, as we know, the cells react to the physical gestures of others. But do monkeys feel sympathy for their fellow primates? Do they understand the frustration of their cage mates? Their happiness? Their guilt? In a word, do monkeys have theory of mind? We are left guessing.

Be that as it may, neuroscientists like Rizzolatti and Iacoboni theorize that human mirror neuron systems have become more advanced than those of our primate cousins, enabling us to have a deep social intelligence that's capable of neurally "inhabiting" the minds of others.

In the years that Rizzolatti's team was first exploring the mirror system, the Italian lab was simultaneously investigating a related set of cells known as canonical neurons, which they found in the same area of the brain. Like mirror neurons, canonical neurons fired whenever monkeys made a grasping gesture. What set canonical neurons apart, however, was that they also spiked in the same grasping pattern when the animal merely *saw* a graspable object.

These twin findings, which played out over years in Rizzolatti's lab, indicated that both humans and monkeys couldn't observe an action *or an object* without our brains unspooling a motor plan of physical interaction. Think about that: if mirror neurons suggest that our brains perceive the actions of others through cellular reenactment, canonical neurons indicate that our perception of inanimate objects is intrinsically bound to our physical interaction with the object.

"In short," Iacoboni writes, "the grasping actions and motor plans necessary to obtain and eat a piece of fruit are inherently linked to our very *understanding* of the fruit. The firing pattern . . . shows clearly that perception and action are not separated in the brain."

Iacoboni writes that subsequent brain-imaging studies have dug even deeper, showing that mirror neurons not only mimic the observed actions of others but also will become active at the very mention of an action. In one fMRI experiment, UCLA researchers asked test subjects to read sentences describing specific gestures associated with eating and grabbing fruit. During the same study, they showed the test subjects videos of people eating and grabbing fruit. They found that when people either read about or watched people eating, areas of the motor cortex associated with mouth and hand movements became active.

In a similar brain-imaging study, Dutch researchers had volunteers listen to someone crunching potato chips or ripping a piece of paper. Researchers watched as neurons in the volunteers' pre-motor cortices sparked to life at the sounds. Test subjects were then given potato chips to eat and sheets of paper to tear. Once again, the same cells that were active at the sounds of eating and tearing became active as the volunteers ate potato chips and tore pieces of paper. Mirror neurons "transform what you see or hear other people do into what you would do yourself," the study's principal investigator, the neuroscientist Christian Keysers, told *Discover* magazine. "You start to really feel what it feels like to do a similar action."

Rizzolatti's research, along with a host of subsequent investigations into the mirror and canonical systems, has had profound implications for more traditional conceptions of brain function, which held that cognition was somehow higher and separate from motor function. These remarkable cells seem to work double duty. They help coordinate physical movement while also contributing to cognition. They indicate that we don't merely conceive of the people and objects around us as distinct from ourselves. Rather, physical interaction with our environment is essential to our perception and understanding. "Planning for motion is the primitive brain construct that lower animals have," said Nicolelis. "We evolved the idea of planning actions way ahead of time, considering the risks. Our ability to plan things out—to strategize—that all started with motion."

The question for Schwartz was how motor cortical cells coded for different objects and actions. Again, Rizzolatti's early studies held valuable clues. Taking brain recordings while his monkeys performed grasping actions, Rizzolatti found that neurons in the pre-motor cortex weren't terribly concerned with whether the monkey was using its left or right hand to grasp an object. The firing patterns for these very different actions weren't too diverse. Where the researchers did see a difference, however, was in the type of grip the monkey employed. The animals' motor neurons would produce one pattern when grasping a small object and a very different one while grasping something large. Rizzolatti's research seemed to indicate that it was the size of the object, not the object itself, that mattered to the brain.

A neuron that fired at the sight of an apple slice would not automatically fire at the sight of a whole apple. On the other hand, the neuron that fired at the sight of an apple slice would also fire at the sight of an orange section. Similarly, the neuron that fired at the sight of an apple would also fire at the sight of an orange. Was the brain simply coding for size and hand grip? After all, we grab a whole apple quite differently than we grab a small slice.

Rizzolatti's early experiments all involved food—oranges, apples, peanuts, and so on. Taking intention into account, was it possible a monkey would perceive all food equally, that is, both apples and oranges are objects to be brought to the mouth and eaten? If so, it could be that the monkey's brain made no distinction between the two objects, representing them both as "food." The difference in motor cortical activation would then involve only the physical details of bringing the apple to its mouth. Should the animal grab the fruit with its whole hand, as it would an entire apple, or should it use a more refined pinch to grasp a slice?

But what about intention? Would the brain activate differently while grabbing an orange section than it would while reaching for a similarly sized nonedible object? That was precisely the question the neuroscientist Leo Fogassi set out to answer a few years ago with a simple reaching task. In the first part of the experiment, Fogassi, who was a member of Rizzolatti's original group, had his monkeys reach for pieces of food to eat. For the second phase, he had the animals reach for a similarly sized object, which they placed in a container by their mouths. The physical actions of grasping to eat and grasping to place in the container were nearly identical, enabling Fogassi and his fellow researchers to concentrate on whether the animals' brains coded differently for edible versus inedible objects. They were testing for intention.

Listening in on the animals' pre-motor cortices, Fogassi found that around 60 percent of the neurons he recorded in the experiment registered different spiking patterns when the animal reached for food versus when the animal reached for the inedible object. To further clarify the data, Fogassi repeated the experiment, this time using only food. In the first phase, he had the monkeys grab food to eat; in the second, he had them grab food to place in the container. Once again, the physical actions were very similar, and Fogassi found that, again, his monkeys coded differently for objects they would eat versus objects they would place in the container. The physical object, Fogassi's research seemed to suggest, mattered less to the brain than how we intend to interact with it.

Fogassi's study didn't stop there. In the next phase, scientists took neural recordings as the monkeys watched researchers perform similar actions, grabbing food to eat in one phase, placing it in a container in the next. Once again, the animals' pre-motor cortices erupted into distinct patterns while observing someone grasp food to eat versus grasping food to place in the container. What's more, the neural activations were similar to when the monkeys performed the actions themselves: the monkeys were coding for intention in the observed researchers, whether to eat or to place the piece of food.

Findings like Fogassi's have prompted neuroscientists and philosophers to theorize that these cells are much more than a mechanism to understand the actions of others. Rather, they are the cellular mechanism we use to divine the mental states (the intentions) of others. And while mirror neurons may serve as the biological underpinning of human empathy, Schwartz believed there was another reason we evolved this sort of embodied consciousness: our own safety. "You're always trying to predict what's going to happen next, so you're watching the way other people move, you're placing their movements in relation to yourself, and you're asking yourself, 'If I was that person, what would I be doing next?' " he said. "Why do you do that? Because the most dangerous things in the world are the people around you. There's a huge evolutionary drive to do that—the same as we are always trying to predict what's going to happen if we touch an object in a certain way."

Schwartz had come a long way from challenging the notion that the motor cortex was merely a control center to move muscles. What he was describing was an essential link between our cognitive representation of the physical world, gross physical movement, and behavior. He was moving toward an idea of embodied consciousness, a consciousness whose very ability to perceive objects and recognize the consciousness of other people is dependent on our own corporeality, our own ability to move. Our physical form and interaction with objects and other people isn't merely the foundation of our consciousness: they are consciousness itself. "Instead of thinking of movement purely in terms of mechanics, you can think of it as behavioral output," he said. "It's not just that the brain is acting as a command center. It is the fact that your physical structure determines your behavior—so the way you are constructed, the way we stand upright on our two legs and are free to move our arms and hands so that they are not supporting the body. That allows us to do a whole range of behaviors that other animals can't do."

Still, observing the brain code objects differently based on intention was a far cry from crafting a brain-computer interface to accommodate this symbolic construction of the world. Nevertheless, Schwartz was increasingly convinced that he would have to take the idea of embodied cognition into account if he was ever to create a brain-computer interface capable of robust and spontaneous movement. He could not think only of moving joint angles and pairing specific neural activity to particular actions.

Perhaps even population vectors and preferred firing directions wouldn't be enough. Perhaps he would have to decode how the embodied brain symbolically perceives the physical world.

After all, he'd done everything right from a technical standpoint. Using population vectors, he had already granted Scheuermann seven degrees of freedom. He knew that, mechanically at least, he'd solved the problem of movement, so how else could he explain Scheuermann's sudden inability to grasp physical objects? Was it merely a bug, or were her motor neurons somehow tied to higher-order thoughts, a complicated relationship that surfaced only when her neural control of the arm began to rival the complexity and grace of natural movement? Like Nicolelis, he would need to record from more neurons.

These questions continued to swirl in Schwartz's mind, but his DARPA funders wanted results. They had him on a strict deadline for his deliverables, and he needed the arm to work. He was essentially the small end of the funnel. After more than six years of planning, untold hours of hard work by hundreds of researchers, and tens of millions of government dollars on the line, it all came down to him. Schwartz had to deliver, so while he continued to puzzle over Scheuermann's inability to grasp, he moved her daily calibration exercises to virtual reality, where Scheuermann used Hector's ghostly avatar to do what she could not in the physical world—grasp and release.

It was a mystery. More mysterious yet: Not long after she switched to the virtual realm, the problem disappeared completely. Scheuermann's brain made some type of adjustment, and she was again able to grasp physical objects.

12. PIANO MAN

"Ah! Rock crushes scissors," Scheuermann crowed, as Hector, the arm's hand in a fist, retracted to its resting position. Draped in a polka-dot fleece, Scheuermann was in the midst of vanquishing her opponent in a game of rock-paper-scissors. Her biological arms—pale, cool, flaccid—rested on a pillow beneath the fleece, while her head, Scheuermann's last outpost of bodily command, lay against her wheelchair's headrest. She hadn't moved on her own for nearly fifteen years, relying instead on the rugged motorized wheelchair she piloted with a shoulder-mounted joystick that she controlled with her chin.

Scheuermann had spent the morning coaxing Hector through a series of elegant arcs, naturalistic reaches, and dexterous grabs, stacking boxes and fitting pegs in a board. Now perched and at the ready, the arm sat off to the side in one of Schwartz's windowless labs.

The average human brain teems with some 100 billion neurons that collectively shimmer with electrochemical consciousness. No one knows how these three pounds of tissue and electricity, awash in a chemical bath of neurotransmitters, result in consciousness, but we do know that delicate yet routine tasks like pinching a few strands of saffron are in fact the collaborative result of thousands of individual neurons that speak to one another. Information moves like sheet lightning across the brain, forming transient patterns of activity that expand and recede as specific clusters of neurons spark other clusters to action.

Scheuermann, by contrast, was using fewer than two hundred cells to control Hector, meaning her broader movements often appeared effortless. But her finer motor skills? Let's just say the system could be a little buggy.

Nevertheless, Scheuermann was already up by one in her game of rock-paper-scissors. Her opponent figured she'd play to her cyborg strengths, urging Hector to flatten for paper instead of forming a more challenging scissors or a fist again for rock.

Scissors beat paper, so that was the plan.

"One, two," she counted as he clapped his fist into his palm, "three!" But as her opponent began to extend his fingers, Hector's network of tiny motors whirred into action. Milliseconds earlier, Scheuermann had imagined extending the middle and index fingers of her right hand. That intention prompted a crackling network of neurons to fire across her brain. The conversation volleyed across her neocortex, but its meaning was most pronounced in the motor strip, where Lewis and Clark, twin listening posts, transmitted the Morse code–like patterns to the beefy gray cables that sprouted hornlike from Scheuermann's skull.

Those cables ferried the electrical pulses of individual neurons to the bank of computers that squatted to the left of Scheuermann's chair. As the signals flowed into the humming cube of black machines, they cascaded across an algorithm that parsed the rat-a-tat-tat of scores of individually firing neurons. The algorithm compared the frequency and amplitude of some neurons with that of others, assessing the emergent firing pattern and transforming it into a series of commands for Hector.

It all happened in a matter of milliseconds—roughly the amount of time it would take a biologically intact nervous system to do the same thing. And as Scheuermann's opponent spread his fingers to form a pair of scissors, Scheuermann, her platinum electrodes, gold wires, signal amplifiers, fiber-optic cables, and powerful computers combining as Hector buzzed forward, matched him digit for digit as the arm unfurled a slender pair of black fingers.

"Tie!" Scheuermann cried as Hector retreated to his neutral crouch.

Down by one, her challenger decided to play it safe. Scheuermann might be a little clumsy with the finer motor tasks, but she was also having a terrific, breakthrough sort of day. An hour earlier, she had begun moving Hector smoothly with a full ten degrees of control—not only a fluid motion that rivaled a natural human arm but also an exponential advance over anything John Donoghue or Miguel Nicolelis had shown to date. She was feeling cocky. She might showboat a little, exhausting the game's repertoire by flattening Hector for paper.

Scissors beat paper, of course, so that was the plan—scissors again. But, fifteen years in a wheelchair had transformed Scheuermann into a fearsome gamer and, as everyone in the room was about to find out, no slouch at rock-paper-scissors, either. "One," she chanted as Hector's motors began to hum, "two . . ."

As Scheuermann called "three," her opponent struck the heel of his right hand into the palm of his left, forming a V with his middle and index fingers. But Scheuermann had anticipated the plan. Hector balled into a fist. "Rock beats scissors!" she cried.

The game was over.

"Next week I'm going to feed myself chocolate," she said, satisfied, as Hector coiled to neutral. "We'll probably need a lot of chocolate."

•

Schwartz had done it. Working for the first time with penetrating electrodes in humans, he'd not merely achieved the study's stated goal of seven degrees of freedom; he'd surpassed it: at their peak, Scheuermann and Hector moved with a full ten degrees of freedom. Schwartz and his colleagues eventually published those findings in 2013 in the British scientific journal *The Lancet* under the title "High-Performance Neuroprosthetic Control by an Individual with Tetraplegia."

By now, Schwartz was prepared for the inevitable media deluge that followed, as *60 Minutes* and other news outlets featured the work. Scheuermann was a little more starstruck as she gave interviews and prepared to be featured on national television. "If you Google my name, I've been in a paper in Switzerland—front page, above the fold," she said. "I've been on the BBC. I've been on the CBC, and all three local papers. I'm just all over the place."

The attention didn't hurt, and Scheuermann, once so despondent, spent the weeks leading up to the *60 Minutes* segment meticulously plotting her social media campaign to drum up interest. On Monday, she'd post a picture on Facebook of her and the correspondent Scott Pelley. "What am I doing here with Scott Pelley?" she planned to post. "Oh yeah! That's because I'm going to be on *60 Minutes*!" She had recruited a few friends as plants. "One of them is going to ask, 'Is it because of your book, *Sharp as a Cucumber*?' And I will say, 'What? You mean this book?' And I'll post a link to my Amazon page." On Wednesday, she'd post a picture of a chocolate bar. On Friday, she'd post her Halloween costume—two figurine-sized ghosts placed atop Lewis and Clark. She was having a ball. "I plan to ride the wave all week," she said. "I'm sure I'll have tons of comments."

Schwartz was more measured. Whereas he'd once dreaded talking to the media, he'd learned to better present his work. "You give them what they want, the usual thing. But then at the end you say, 'But what's really cool is that it's allowing us to probe the brain and learn things that we weren't able to do before,'" he said. "I mean, it's great to watch our subject move. It's very pleasing. It's emotional. But the thing that drives me is the science. You know, we got it to work. Let's move on."

But there was no denying it: Schwartz was pleased with the study. He'd spent his entire career waiting for just this moment. Now that it was here, he was less concerned with the public attention than he was with the small group of researchers he counted as colleagues.

Schwartz was finally on top. Donoghue and Nicolelis might have been first, but neither of them had achieved anything approaching his elegance or complexity of control. "Look at that," he said, watching a video of Scheuermann and Hector. "That is precise. Impressive, huh? That's the real thing. I'm not shooting smoke, man." He would continue to work with DARPA, trying to recruit new patients the program planned to implant with wireless interfaces and endow with a digital sense of touch. But Schwartz felt that in some essential way he'd done it. He'd unlocked part of the mystery of movement. "This paper conclusively shows that that is a reality. It's validated that concept," he said. "There's more work to be done and everything, but I think that just about everybody would be convinced that it can happen—that it will happen."

The effort, some thirty-five years in the making, had done well by him. Schwartz was undeniably a star in his rarefied realm. He had appeared in nearly every major media outlet. He traveled the world to present his work at conferences. DARPA was thrilled with his progress, and he now boasted one of the biggest, most advanced neuroscience labs in the world.

Still, other mysteries remained, and Schwartz was growing tired of the BCI race. "Maybe I'm just getting older, but I just feel more secure," he said. "I don't want to overblow it, but it's like in an athletic thing. You've won some sort of contest. I kind of feel like I've proven myself, and that's satisfying."

He now longed for the simpler days when he was first starting out and built his own task chair to start a monkey lab. Back then, he worked with one lab tech and one monkey. He had reconfigured an early touch screen from an ATM to use in experiments. He built all of his own equipment. He did all of his own programming. He trained his own monkey. He hadn't had to teach, and he didn't supervise anyone. It was pure research, just Schwartz and his monkey. "It was fantastic, completely hands-on," he said.

"When we finished, the technician and I, we got a bottle of champagne and had our picture taken with that monkey. My first monkey."

Now, after years spent chasing his neuroprosthetic dream, Schwartz wanted to return to those roots, diving deeper into the brain's mysteries.

He wanted to discover what he believed was the holy grail of neuroscience: namely, the underlying principles that cause a cell to produce an action potential. Any given neuron is connected to some ten thousand other cells, many of which are firing and releasing neurotransmitters to the receiving cell. If enough of those surrounding cells are active and their activity is intense enough, the receiving cell will generate an action potential.

But why is that?

"Are they different neurons communicating every time? Is every neuron equal? Or are some guys special?" he asked. "Is it always the same synapses? Do their inputs have preferred directions, and the ones that fire together in a certain way are the ones that make these guys fire? Or is there some other rule, an algebraic sum?"

It's no idle question. As a neuron changes allegiances, becoming more associated with some cells than with others, the neuroplastic brain is not merely changing its physiology. It is learning, changing its physical makeup to alter its behavior. Schwartz had shown this process again and again with BCIs, prompting subjects to alter their brain's firing patterns to better control a device. Now he wanted to go deeper. He didn't want to merely observe cells change: he wanted to investigate the fundamental synaptic mechanism that underlies that change. He wanted to know why a cell fires. "And I have the experiment now that I can do to answer that question," he said.

It would take just one monkey. Schwartz could do it in his lab's basement, mapping a neuron's synaptic connections to investigate how those synapses shifted their behavior, prompting a change in the receiving cell. Would the cell stop generating action potentials when a specific synapse stopped releasing neurotransmitters? Was it simply a numbers game? Or were there essential synaptic combinations without which the cell will not fire?

Schwartz didn't want merely to watch the brain learn anymore. He wanted to understand this most fundamental subcellular mechanism that makes not only learning but also movement, emotions, reason, perception, ideas, fantasies, and dreams possible. "Maybe it won't pan out," he said, "but I'd sure like to give it a shot. I mean, we have no idea what makes a neuron fire, and that's at the root of everything."

EPILOGUE

Each year, scientists from around the country meet for the Society for Neuroscience conference, the field's largest meeting, where researchers unveil their latest findings and size up the competition. Held in New Orleans in 2012, the SfN conference took place before Schwartz had published his findings. His lab had been criticized earlier for talking to the press about their work with Hemmes before publishing their results in an academic journal. But by then, Schwartz had already achieved ten degrees of freedom with Scheuermann, and he roamed the halls of the cavernous Ernest N. Morial Convention Center, showing a select group of colleagues videos on his smartphone.

While the Schwartz group hadn't formally published its findings, it presented a poster of its work, where Donoghue, Hochberg, Leuthardt, Moran, and the rest of the field could see the preliminary results. Conspicuously absent from the meeting was Nicolelis. "I only go to SfN every two years because I find it so repetitive," he said later. "It's not worth the taxpayers' money to go."

Nevertheless, just as Hochberg and Donoghue had been the stars of the Neural Interfaces Conference back in Salt Lake City, the small group of BCI researchers was abuzz over Schwartz's research.

On the third night of the conference, many of the field's brightest stars gathered for a party in the courtyard of the House of Blues, where the drink list, titled "Beverage Computer Interfaces," offered everything from "Non-invasive beverages" like seltzer and Diet Coke to neuro-takes on drink classics like the "Eric Lemonhardt" (vodka and lemon), the "Cosmopolitan Donoghue" (a cosmopolitan), and "Nicho's Network" (a whiskey sour).

"Everyone here is a competitor," Hochberg said in a brief moment of candor away from the crowd. Nevertheless, he and the rest of the researchers were all in a festive mood that night. Schwartz might have struck hard with the Scheuermann research, but Hochberg, dressed in a roomy pin-striped suit, was carrying in his pocket the future of the field: a matchbook-sized prototype of a wireless telemetry device. "It should be a nice continuation of the field," he allowed.

By then, Leuthardt was well on his way with Neurolutions, the BCI-based company he'd started with Moran and Schalk. They were in the midst of fabricating the IpsiHand, the orthotic device they hoped would enable BCI stroke patients to regain function. Toggling as always between his surgery practice and his research, Leuthardt had added fund-raising to his repertoire, meeting with potential investors and medical device manufacturers. Still, he was in the midst of getting the study approved. Medical school administrators were hesitant to sign off on an academic study that could potentially benefit a private company. "Conflict of interest affects everybody," he said. "It's really this war of two cultures: if you were to divest yourself of all financial interest in the company, investors just couldn't understand that."

The university would eventually sign off on Leuthardt's study, whose preliminary results have been promising. In the meantime, Brookman, who returned home to Tulsa after the surgery, was slowly recuperating, working with speech and physical therapists to regain some of the function he'd lost after the second operation. His was a long and halting recovery, and he still struggled to control his Tourette's syndrome and bipolar disorder. Nevertheless, the surgery itself had been an unequivocal success. "After the surgery, it's like, boom, I stopped having seizures immediately," he said. More than a year after his procedure, Brookman had not suffered a single epileptic episode.

Leuthardt had since worked with several other research subjects, and his lab presented posters of their work using ECoG BCIs to interpret speech and re-create motor activity. Still, Leuthardt was running up against the same issues Schwartz and the rest of the field encountered with BCIs. Sure, they could re-create movement, but was that enough? Did they really understand the underlying mechanisms?

"I've seen a lot of what I call information transfer functions," he'd said at lunch earlier that day. "I take brain activity, and I predict finger movements, but it's an informational transfer: you put in the variables, and the machine-learning algorithm spits out its ability to predict. But it doesn't tell you what's going on. It's a black box."

At the party, however, these larger concerns were put away for the night. Hochberg gave a speech and presented an award, while Donoghue and Schwartz mingled with their respective labs. This uneasy fellowship had called a truce for the night, and as the scientists slowly retired to their hotels, Gerwin Schalk lingered in the doorway. The former football player was dressed in a red John Lennon T-shirt, stabbing at the air as he talked with fellow researchers. Sure, their BCIs re-created movement, but to what end? How would their results be translated into a marketable device? "They're just so enthralled with their neurons," he said. "But that's only the first 10 percent of the work."

NOTES

ACKNOWLEDGMENTS

INDEX

NOTES

1. BYPASSING THE BODY

Electrocorticography, or ECoG, was first developed in the 1950s by the neurosurgeon Wilder Penfield and his colleague the neurophysiologist Herbert Jasper. They developed ECoG as part of the so-called Montreal procedure, a surgical technique that combined the principles of neurophysiology with neurosurgery and used electrodes not only to identify epileptic tissue but also to create functional brain maps of motor and sensory areas to be avoided during surgery. A discussion of their work can be found in Penfield and Jasper's seminal text, *Epilepsy and the Functional Anatomy of the Human Brain* (London: J. and A. Churchill, 1954).

Philip Kennedy describes his early work with implanting neurotrophic electrodes to give people control of computers in a 1998 paper in *NeuroReport.* Philip Kennedy and Roy Bakay, "Restoration of Neural Output from a Paralyzed Patient by a Direct Brain Connection," *NeuroReport,* June 1, 1998, 1707–11. Similarly, Niels Birbaumer described his work using an EEG apparatus to enable research subjects to control basic word-processing software in 1999. Niels Birbaumer et al., "A Spelling Device for the Paralysed," *Nature,* March 25, 1999, 297–98.

A vivid retelling of Nicolelis's early research with a monkey named Aurora can be found in his autobiographical account of his scientific career, *Beyond Boundaries* (New York: Henry Holt, 2011), 125–55. Nicolelis first had the computer send Aurora's brain signals to a robotic arm in the next room as she used a joystick to play a video game. Later, Nicolelis writes, they took the joystick away, linking control of the video game directly to the robot arm she controlled by her brain. "Aurora was now playing her video game just by thinking," he writes (p. 153). "No need to use her own arms anymore. Her brain activity, free of her body and self-sufficient, was carrying out, across laboratory walls, every bit of the burden generated by her voluntary will."

In 2010, DARPA formed the Reliable Neural-Interface Technology

(RE-NET) program with an aim toward producing a more dependable means of extracting neural information. According to the DARPA program's Web site, http://www.darpa.mil/Our_Work/BTO/Programs/Reliable_Neural-Interface_Technology_RE_NET.aspx, "RE-NET seeks to develop the technologies needed to reliably extract information from the nervous system, and to do so at a scale and rate necessary to control many degree-of-freedom (DOF) machines, such as high-performance prosthetic limbs. Prior to the DARPA RE-NET program, all existing methods to extract neural control signals were inadequate for amputees to control high-performance prostheses, either because the level of extracted information was too low or the functional lifetime was too short."

For a discussion of the problems and solutions in EEG signal interference, I found Dean J. Krusienski, Dennis J. McFarland, and Jose C. Principe, "BCI Processing: Signal Extraction," in *Brain-Computer Interfaces: Principles and Practice*, ed. Jonathan Wolpaw and Elizabeth Winter Wolpaw (New York: Oxford University Press, 2012), 123–46, quite helpful.

Writing for *Discover* magazine, the journalist Adam Piore explored the efforts of one of Leuthardt's coinvestigators, Gerwin Schalk, as it related to the army's efforts to build a "thought helmet." Adam Piore, "The Army's Bold Plan to Turn Soldiers into Telepaths," *Discover*, April 2011.

For further discussion on memory prostheses, see Theodore Berger et al., "A Cortical Neural Prosthesis for Restoring and Enhancing Memory," *Journal of Neural Engineering* 8, no. 4 (2011): 046017.

For more on the details of Nicolelis's controversial work endowing animals with a "sixth sense," enabling them to perceive infrared light through a sense of touch, see Eric E. Thomson, Rafael Carra, and Miguel Nicolelis, "Perceiving Invisible Light Through a Somatosensory Cortical Prosthesis," *Nature Communications* 4, article no. 1482 (2013).

Ramez Naam's *More Than Human* (New York: Broadway Books, 2005) gives a sense of the sometimes-breathless enthusiasm many futurists have for the field of neuroprosthetics. While discussing the possibility of a brain-to-brain interface, Naam writes, "Rather than having to guess what your spouse or child is feeling, you would simply be sensing it via the wireless link between your brains. If you wanted to sense other people's feelings less, you could choose to turn down the volume, reducing the strength of the signal that your neural interface sent into your empathy centers. Software could even decide which people's feelings to send into your empathy centers and which not to. The end result might be just like having an unusually keen sense of how others are feeling" (p. 196).

The American Speech-Language-Hearing Association's technical report (http://www.asha.org/policy/TR2004-00041/#sec1.2) provides a succinct but thorough overview of the history and technology of cochlear implants. Similarly, I found Vittorio A. Sironi, "Origin and Evolution of Deep Brain Stimulation," *Frontiers in Integrative Neuroscience*, August 18, 2011, enlightening on the history and future potential of deep-brain stimulation.

My reading of Plato's *Phaedrus*, trans. Benjamin Jowett (Boston: Actonian Press, 2010), is deeply influenced by the work of Jacques Derrida, particularly "The Pharmakon" in *Dissemination*, trans. Barbara Johnson (Chicago: University of Chicago Press, 1981), which alerts readers to Plato's fundamental distrust of technology. For more on the role technology plays in our conception of what it means to be human, see Andy Clark, *Natural-Born Cyborgs* (New York: Oxford University Press, 2004).

For more about the neural representation of the left hand in violinists, see Thomas Elbert et al., "Increased Cortical Representation of the Fingers of the Left Hand in String Players," *Science* 270 (1995): 305–9. Similarly, Angelo Maravita and Atsushi Iriki, "Tools for the Body (Schema)," *Trends in Cognitive Science* 8, no. 2 (2004): 79–86, discusses in greater depth the observed cortical reorganization that attends prolonged tool use.

2. DARPA HARD

I found Alan J. Thurston, "Paré and Prosthetics: The Early History of Artificial Limbs," *ANZ Journal of Surgery* 77, no. 12 (2007): 1114–19, particularly insightful on the subject of early upper-limb prostheses, and I relay many of the details the article contains here. Along those same lines, Philippe Hernigou, "Ambroise Paré IV: The Early History of Artificial Limbs (from Robotic to Prostheses)," *International Orthopaedics* 37, no. 6 (2013): 1195–97, is fascinating. A more recent history may be found in Thelma L. Wellerson, "Historical Development of Upper Extremity Prosthetics," *Orthopedic and Prosthetic Appliance Journal* 11, no. 3 (1957): 73–77, which also provides some of the details I reference in the text. For a closer look at the issues regarding early myoelectric upper-limb prosthetics, I found Roy Wirta, Donald Taylor, and F. Ray Finley, "Pattern-Recognition Arm Prosthesis: A Historical Perspective—a Final Report," *Bulletin of Prosthetics Research* (Fall 1978): 8–35, quite helpful.

I relied on Craig L. Taylor and Robert J. Schwarz, "The Anatomy and Mechanics of the Human Hand," *Artificial Limbs* 2, no. 2 (1955): 22–35, for my discussion of the hand's anatomical complexity.

I relied on several articles in the press as well as interviews for the section on Dean Kamen's effort to develop the DEKA arm. I found Sarah Adee, "Dean Kamen's 'Luke Arm' Prosthesis Readies for Clinical Trials," *IEEE Spectrum*, February 2008, http://spectrum.ieee.org/biomedical/bionics/dean-kamens-luke-arm-prosthesis-readies-for-clinical-trials, particularly good.

Todd A. Kuiken et al., "Targeted Muscle Reinnervation for Real-Time Myoelectric Control of Multifunction Artificial Arms," *Journal of the American Medical Association* 301, no. 6 (2009): 619–28, gives a detailed account of the science behind their pioneering method of targeted muscle reinnervation.

For an exhaustive discussion of the scientific and technical challenges that accompanied the development of APL's modular prosthetic limb, readers should consult a suite of APL articles published in the *Johns Hopkins APL Technical Digest* 30, no. 3 (2011): 182–266: Dexter G. Smith and John D. Bigelow, "Biomedicine: Revolutionizing Prosthetics—Guest Editors' Introduction," 182–85; Stuart Harshbarger, John Bigelow, and James Burck, "Revolutionizing Prosthetics: Systems Engineering Challenges and Opportunities," 186–97; Robert S. Armiger et al., "A Real-Time Virtual Integration Environment for Neuroprosthetics and Rehabilitation," 198–206; Matthew S. Johannes, et al., "An Overview of the Developmental Process for the Modular Prosthetic Limb," 207–16; Michael M. Bridges, Matthew P. Para, and Michael J. Mashner, "Control System Architecture for the Modular Prosthetic Limb," 217–22; Todd J. Levy and James D. Beaty, "Revolutionizing Prosthetics: Neuroscience Framework," 223–29; Francesco V. Tenore and R. Jacob Vogelstein, "Revolutionizing Prosthetics: Devices for Neural Integration," 230–39; Courtney W. Moran, "Revolutionizing Prosthetics 2009 Modular Prosthetic Limb–Body Interface: Overview of the Prosthetic Socket Development," 240–49; Paul J. Biermann, "The Cosmesis: A Social and Functional Interface," 250–55; and Mark A. Hinton et al., "Advanced Explosive Ordnance Disposal Robotic System (AEODRS): A Common Architecture Revolution," 256–66.

For more on the military's funding of brain research, Jonathan Moreno's *Mind Wars* (New York: Bellevue Literary Press, 2012) is a provocative and at times startling account of the military's attempts to "weaponize" the human brain. Moreno, an ethics professor at the University of Pennsylvania, writes about numerous historical, recent, and ongoing government programs, detailing efforts to reduce soldiers' need for sleep and to enhance cognition.

3. MONKEY MAN

Jon Mukand, *The Man with the Bionic Brain* (Chicago: Chicago Review Press, 2012), has been an invaluable resource, affording an inside look at the BrainGate pilot study and Matthew Nagle's personal experience and concerns following the knife attack. I have relied on some of his reflections in recounting the attack that left Nagle paralyzed. Ingrid Wickelgren, "Tapping the Mind," *Science* 299 (January 2003): 496–99, gives a good overview of early BCI researchers and government funding.

For an account in the popular press of the Cyberkinetics/BrainGate pilot study, Richard Martin, "Mind Control," *Wired*, March 2005, provides many interesting details about the research and personalities behind it. Similarly, Leander Kahney named Nagle the world's first "neuro-cybernaut" in "Biggest Discoveries of 2005," *Wired*, December 2005, http://archive.wired.com/science/discoveries/news/2005/12/69909?currentPage=all. In the article, Kahney wrote, "A 'Braingate' chip implanted in Nagle's motor cortex allows him to reach out and grasp objects by thinking about moving his own paralyzed hand. Nagle's neuro-cybernetic interface also allows him to control the lights, TV and a computer. 'My mother was scared of what might happen, but what else can they do to me?' Nagle said. 'I was in a corner, and I had to come out fighting.' "

For a more formal account of the BrainGate pilot study, see Leigh R. Hochberg et al., "Neuronal Ensemble Control of Prosthetic Devices by a Human with Tetraplegia," *Nature*, July 13, 2006, 164–71. Donoghue's quotation "If your brain can do it, we can tap into it" appeared in Andrew Pollack, "Paralyzed Man Uses Thoughts to Move a Cursor," *New York Times*, July 13, 2006. His quotation about the "dawn of the age of neurotechnology" appears in an unattributed article titled "Brain Chip Heralds Neurotech Dawn" posted to the CNN Web site, July 17, 2006, http://www.cnn.com/2006/TECH/science/07/17/braingate.donoghue/index.html?iref=newssearch.

For an engaging discussion on the neuroscientist Daniel Wolpert's theories about the brain's evolutionary link to muscular control, see his talk "The Real Reason for Brains" on the TED Web site, July 2011, http://www.ted.com/talks/daniel_wolpert_the_real_reason_for_brains/transcript?language=en.

Marcel Adam Just et al., "A Neurosemantic Theory of Concrete Noun Representation Based on the Underlying Brain Codes," *PLoS ONE*, January 13, 2010, doi: 10.1371/journal.pone.0008622, gives an account of their research into the brain's representation of nouns. Just's comment that we are "fundamentally perceivers and actors" comes from the press release "Carnegie Mellon Computer Model Reveals How Brain Represents Meaning" provided by Carnegie Mellon University, May 29, 2008, http://www.cmu.edu/news/archive/2008/May/may29_brainmeaning.shtml.

For the discussion of Edward Evarts's work and career, I relied on William Thach's biographical memoir, "Edward Vaughan Evarts, 1926–1985," in *Biographical Memoirs* (Washington, D.C.: National Academy Press, 2000), vol. 78.

Apostolos P. Georgopoulos, Andrew B. Schwartz, and Ronald E. Kettner, "Neuronal Population Coding of Movement Direction," *Science* 233, no. 4771 (1986): 1416–19, describes their early research into the directional tuning of neurons. Similarly, Schwartz's groundbreaking work giving a monkey direct neural control over a neuroprosthetic device is described in Dawn M. Taylor, Stephen I. Helms Tillery, and Andrew B. Schwartz, "Direct Cortical Control of 3D Neuroprosthetic Devices," *Science* 296, no. 5574 (2002): 1829–32.

Miguel Nicolelis describes his work with the owl monkey Belle in his book *Beyond Boundaries*, 137–45, and again in Johan Wessberg et al., "Real-Time Prediction of Hand Trajectory by Ensembles of Cortical Neurons in Primates," *Nature*, November 16, 2000, 361–65.

Schwartz disputes Nicolelis's claim that he was the first researcher to predict movement by decoding neural activity, pointing to Robert E. Isaacs, Douglas J. Weber, and Andrew B. Schwartz, "Work Toward Real-Time Control of a Cortical Neural Prosthesis," *IEEE Transactions in Rehabilitation Engineering* 8, no. 2 (June 2000): 196–98.

4. BAD CODE

For a scholarly look at the McGurk effect, I found Kaisa Tiippana, "What Is the McGurk Effect?," *Frontiers of Psychology* 5, no. 725 (2014), doi:10.3389/fpsyg.2014.00725, helpful. To experience the McGurk effect personally, readers should watch the demonstration on BBC Two's *Horizon* (2010). Although the video is no longer available on the main site, readers can still view it on the BBC's YouTube channel, accessed November 10, 2010, https://www.youtube.com/watch?v=G-lN8vWm3m0. It's astonishing.

The National Institute of Neurological Disorders and Stroke has a very good primer on the causes, types, and treatments of epilepsy, accessed February 4, 2015, http://www.ninds.nih.gov/disorders/epilepsy/detail_epilepsy.htm.

For more about how the brain processes homonyms, see Andrea J. R. Balthasar, Walter Huber, and Susanne Weis, "A Supramodal Brain Substrate of Word Form Processing: An fMRI Study on Homonym Finding with Auditory and Visual Input," *Brain Research* 2, no. 1410 (2011): 48–63.

5. SCREW THE RATS!

My discussion of Darwinian evolution is heavily influenced by the writings of the evolutionary anthropologist David Sloan Wilson. His book *Evolution for Everyone: How Darwin's Theory Can Change the Way We Think About Our Lives* (New York: Random House, 2007) is a superb resource. Similarly, I relied heavily on writings of Mark Pagel in my discussion of genetics and culture. His book *Wired for Culture: Origins of the Human Social Mind* (New York: W. W. Norton, 2012) is an elegant treatise on the role culture has played in shaping modern humanity. I find his ideas both provocative and profound, and I am deeply indebted to his insights in forming my own views on the role that culture, genetics, and technology (and particularly emerging neurotechnologies) play from an evolutionary standpoint in shaping modern humans. Similarly, the work of the evolutionary biologist Richard Dawkins—and particularly his book *The Selfish Gene* (New York: Oxford University Press, 1990)—has deeply influenced my thinking on evolution and genetics.

The coining of the term "cyborg" dates back to Manfred Clynes and Nathan Kline, "Cyborgs and Space," *Astronautics*, September 1960, 26–27, 74–76. The journalist Geoffrey Pond described Clynes's Rockland laboratory in "Young Scientist Leads Two Lives," *New York Times*, March 20, 1960. Similarly, the journalist Alexis Madrigal interviewed Clynes on the fiftieth anniversary of the *Astronautics* article. The interview, titled "The Man Who First Said 'Cyborg,' 50 Years Later," *Atlantic*, September 30, 2010, http://www.theatlantic.com/technology/archive/2010/09/the-man-who-first-said-cyborg-50-years-later/63821/, draws a comparison between the idea of the cyborg and the computer of average transients (CAT) machine. Discussing the machine's bypassing of conscious thought to measure evoked neural responses, Clynes said, "It was a way of finding the needle in the haystack." He added, "Let's say you had a light stimulus of a certain color and you wanted to see the influence of looking at that color had on the electrical activity of the brain. You presented the color a few times and averaged the result."

The functioning and the utility of BCI2000 software are described in Gerwin Schalk et al., "BCI2000: A General-Purpose Brain-Computer Interface (BCI) System," *IEEE Transactions on Biomedical Engineering* 51, no. 6 (June 2004): 1034–43.

The role of high-frequency brain signals in ECoG is explored in Daniel Moran, "Evolution of Brain-Computer Interface: Action Potentials, Local Field Potentials, and Electrocorticograms," *Current Opinion in Neurobiology* 20, no. 6 (2010): 741–45.

A description of Nicolelis's early rat BCIs may be found in John K. Chapin et al., "Real-Time Control of a Robot Arm Using Simultaneously Recorded Neurons in the Motor Cortex," *Nature Neuroscience* 2, no. 7 (July 1999): 664–70.

Similarly, details of the first BCI to use ECoG signals are recounted in Eric Leuthardt et al., "A Brain Computer Interface Using Electrocorticographic Signals in Humans," *Journal of Neural Engineering* 1, no. 2 (June 2004): 63–71.

The stroke statistics I cite come from the Web site for the Centers for Disease Control and Prevention, May 7, 2014, http://www.cdc.gov/stroke/facts.htm. For more about ECoG BCI's potential to detect neural patterns associated with ipsilateral movement, see Kimberly J. Wisneski et al., "Unique Cortical Physiology Associated with Ipsilateral Hand Movements and Neuroprosthetic Implications," *Stroke* 39, no. 12 (December 2008): 3351–59.

6. THE BACKUP PLAN

For more on Rizzolatti's original research, see Giuseppe di Pellegrino et al., "Understanding Motor Events: A Neurophysiological Study," *Experimental Brain Research* 91, no. 1 (1992): 176–80. Also helpful is Giacomo Rizzolatti et al., "Premotor Cortex and the Recognition of Motor Actions," *Cognitive Brain Research* 3, no. 2 (March 1996): 131–41. In addition, Vittorio Gallese et al., "Action Recognition in the Premotor Cortex," *Brain* 119, no. 2 (April 1996): 593–609, is that journal's most cited article.

Marco Iacoboni's lucid book, *Mirroring People* (New York: Farrar, Straus and Giroux, 2008), has also been essential in formulating my thoughts on the mirror-neuron system.

Again, for a formal discussion of Nagle's work with the BrainGate clinical trial, see Hochberg et al., "Neuronal Ensemble Control of Prosthetic Devices by a Human with Tetraplegia." Interested readers may also find videos of Nagle's performance on the *Nature* Web site, http://www.nature.com/nature/focus/brain/experiments/index.html.

Schwartz's groundbreaking work, using a BCI for self-feeding, is detailed in Meel Velliste et al., "Cortical Control of a Prosthetic Arm for Self-Feeding," *Nature*, June 19, 2008, 1098–101. Readers interested in seeing Schwartz's monkeys in action can also find videos of the research on the Web site for Schwartz's lab at the University of Pittsburgh, http://motorlab.neurobio.pitt.edu/multimedia.php.

For a fuller discussion of preferred neuronal firing directions, see Georgopoulos, Schwartz, and Kettner, "Neuronal Population Coding of Movement Direction." By the time Schwartz completed the 2008 self-feeding study, his mentee Daniel Moran had already found success with ECoG. A description of Moran's early work may be found in Leuthardt et al., "a Brain Computer Interface Using Electrocorticographic Signals in Humans."

For an academic account of Wei Wang's research with Tim Hemmes, see Wei Wang et al., "An Electrocorticographic Brain Interface in an Individual with Tetraplegia," *PLoS ONE*, February 6, 2013, doi:10.1371/journal.pone.0055344. Readers interested in seeing Hemmes control the robotic limb during the study can also find videos of the work on the University of Pittsburgh Medical Center/University of Pittsburgh School of the Health Sciences' Web site, http://www.upmc.com/media/Pages/video.aspx?vcat=543%3b%23e2f9d53a-9732-4bb6-a192-07bd74771b65%7cBrain+Computer+Interface+Research.

7. FEELING THE LIGHT

Readers interested in exploring the early work of Miguel Nicolelis and John Chapin should consult Chapin et al.'s seminal article, "Real-Time Control of a Robot Arm Using Simultaneously Recorded Neurons in the Motor Cortex." Wessberg et al., "Real-Time Prediction of Hand Trajectory by Ensembles of Cortical Neurons in Primates," provides additional insight into this early work.

For a formal account of Cherry's study, see Peter J. Ifft et al., "A Brain-Machine Interface Enables Bimanual Arm Movements in Monkeys," *Science Translational Medicine* 5, no. 210 (November 2013): 210ra154.

Similarly, an account of the Nicolelis Lab's efforts to develop wireless BCI may be found in David A. Schwarz et al., "Chronic, Wireless Recordings of Large-Scale Brain Activity in Freely Moving Rhesus Monkeys," *Nature Methods* 11, no. 6 (2014): 670–76.

Robert Wurtz gives a comprehensive telling of the scientific significance of David Hubel and Torsten Wiesel's work in "Recounting the Impact of Hubel and Wiesel," *Journal of Physiology* 587, no. 12 (June 2009): 2817–23. For a taste of their original research, see David Hubel and Torsten Wiesel, "Receptive Fields of Single Neurones in the Cat's Striate Cortex," *Journal of Physiology* 148, no. 3 (October 1959): 574–91, as well as David Hubel and Torsten Wiesel, "Receptive Fields, Binocular Interaction, and Functional Architecture in the Cat's Visual Cortex," *Journal of Physiology* 160, no. 1 (1962): 106–54.

Miguel Nicolelis provides James McIlwain's quotation about how studying an individual neuron can prompt researchers to inflate its importance in *Beyond Boundaries*, 81. Similarly, I rely on Nicolelis's book, along with interviews, for details of his early years as a medical student in Brazil (Nicolelis, *Beyond Boundar-*

ies, 89), as well as his interest (along with John Chapin) in multicellular recordings and their early experiments to understand the relationship between the snout region and the sensory cortex in rodents (Nicolelis, *Beyond Boundaries*, 93–124). A more academic account of that research may be found in Miguel A. L. Nicolelis et al., "Sensorimotor Encoding by Synchronous Neural Ensemble Activity at Multiple Levels of the Somatosensory System," *Science* 268, no. 5215 (June 1995): 1353–58. Similarly, Chapin et al., "Real-Time Control of a Robot Arm Using Simultaneously Recorded Neurons in the Motor Cortex," recounts their seminal work enabling a rodent to control a feeding lever with its thoughts.

Again, Nicolelis gives an account of his work with the owl monkey Belle, linking her brain to robotic limbs that were hundreds of miles away, in his book *Beyond Boundaries*, 137–45. The researcher also describes the study in Wessberg et al., "Real-Time Prediction of Hand Trajectory by Ensembles of Cortical Neurons in Primates."

Nicolelis describes his experiment to grant a monkey neural control over a Japanese walking robot in *Beyond Boundaries*, 182–94. Interested readers can find a video of the transpacific BCI on the Nicolelis Lab's Web site, http://www.nicolelislab.net/?page_id=79.

A fuller account of Nicolelis's experiments with rats' perception of infrared light can be found in Thomson, Carra, and Nicolelis, "Perceiving Invisible Light Through a Somatosensory Cortical Prosthesis." Interested readers can find video of the infrared experiment on the BeyondBoundariesBook YouTube channel, https://www.youtube.com/watch?v=nsniwzap2qE.

A scholarly account of the scientist's brain-to-brain BCI work can be found in Miguel Pais-Vieira et al., "A Brain-to-Brain Interface for Real-Time Sharing of Sensorimotor Information," *Scientific Reports*, February 28, 2013, doi:10.1038/srep013192013. Once again, a video documenting part of the experiment can be found on the BeyondBoundariesBook YouTube channel, https://www.youtube.com/watch?v=nNuntbrwXsM.

8. CYBERKINETICS

Tony Judt's moving essay about his progressive paralysis, "Night," appeared in *The New York Review of Books*, January 14, 2010. As noted in the text, there are only a handful of accounts from people who are fully locked in. This makes Jean-Dominique Bauby's lyrical memoir, *The Diving Bell and the Butterfly* (New York: Vintage, 1998), all the more astonishing. The book was later made into a film of the same name directed by Julian Schnabel (2007).

I rely on the journalist Jessica Benko's article "The Electric Mind," *Atavist*, no. 15 (May 2012), for some of the details of Cathy Hutchinson's life and her participation in the BrainGate study. Similarly, Bob Veillette's life and participation in the study were the subject of a series of articles by Tracey O'Shaughnessy in the *Waterbury Republican-American*: "A Locked-In Love," May 20, 2012;

"Grasping for Hope," May 21, 2012; "Now What?," May 22, 2012; "Parallel Struggles," May 23, 2012; and "A Symphony of the Mind," May 24, 2012.

Martin, "Mind Control," provides some details of Donoghue's early life, his meeting with Richard Normann, and his work with Matthew Nagle.

A scholarly account of Donoghue's early work with the Utah array to determine its utility and stability can be found in Mijail D. Serruya et al., "Instant Neural Control of a Movement Signal," *Nature*, March 14, 2002, 141–42.

Donoghue's quotation "You can substitute brain control for hand control, basically," appeared in Andrew Pollack, "With Tiny Brain Implants, Just Thinking May Make It So," *New York Times*, April 13, 2004.

Nicolelis's early work recording simultaneously from multiple electrodes is recounted in Miguel A. L. Nicolelis et al., "Dynamic and Distributed Properties of Many-Neuron Ensembles in the Ventral Posterior Medial Thalamus of Awake Rats," *Proceedings of the National Academies of Sciences of the United States of America*, March 15, 1993, 2212–16.

An account of Cyberkinetics' reverse merger with Trafalgar Ventures appears in the staff-generated article "Cyberkinetics 'Goes Public' in Reverse Merger," *Boston Business Journal*, October 4, 2004. Donoghue's quotation "It's Luke Skywalker" appears in Kevin Maney, "Scientists Gingerly Tap into Brain's Power," *USA Today*, October 10, 2004. Friehs's quotation that the technology was "almost unbelievable" appears in Simon Hooper, "Brain Chip Offers Hope for Paralyzed," CNN Web site, October 21, 2004, http://www.cnn.com/2004/TECH/10/20/explorers.braingate/.

As noted earlier, in addition to interviews, I rely on Mukand's *Man with the Bionic Brain* for several details of the BrainGate trial, including internally circulated e-mails and Mukand's impressions of Nagle's life as a research subject.

A history of Cyberkinetics' filings with the SEC is available at the NASDAQ Web site, http://www.nasdaq.com/markets/spos/filing.ashx?filingid=4816401. Nagle's quotation that he "can bring the cursor just about anywhere" appeared in Martin, "Mind Control." Nagle's testimony from Cirignano's trial comes from Johanna Seltz, "Paralyzed Ex-Weymouth Football Star Has New Goal," *Boston Globe*, April 2, 2006.

Again, for an official account of Nagle's BrainGate study, readers should consult Hochberg et al., "Neuronal Ensemble Control of Prosthetic Devices by a Human with Tetraplegia." Donoghue's quotation "If your brain can do it, we can tap into it" appeared in Pollack, "Paralyzed Man Uses Thoughts to Move a Cursor." Readers interested in seeing videos of Nagle's performance should consult *Nature*'s Web site, where supplementary information accompanies Donoghue's original article, http://www.nature.com/nature/journal/v442/n7099/suppinfo/nature04970.html.

The editorial "Is the University-Industrial Complex out of Control?" appears in *Nature*, January 11, 2001, 119.

The market estimations for neural stimulation and global spinal cord injury are gleaned from the 2006 press release "Cyberkinetics Neurotechnology Inc. Agrees to Acquire Andara Life Science Inc." that accompanied the acquisition and was distributed on behalf of Andara by Purdue Research Foundation.

For more on Cyberkinetics, Andara, and the company's sale of its assets to NeuroMetrix, see Scott Kirsner, "Even with Brilliant Idea and Deep Pockets, Risks High for Start-Ups," *Boston Globe*, April 12, 2009. For more on Jeff Stibel's purchase of the BrainGate trademark and some of its patent claims, see Scott Kirsner, "CyberKinetics' Brain-to-Computer Interface Gets a Second Chance," *Innovation Economy* (blog), *Boston Globe*, August 12, 2009, http://www.boston.com/business/technology/innoeco/2009/08/cyberkinetics_braintocomputer.html.

For an academic account of Donoghue's research with Hutchinson and Veillette, see Leigh R. Hochberg et al., "Reach and Grasp by People with Tetraplegia Using a Neurally Controlled Robotic Arm," *Nature*, May 17, 2012, 372–75. Interested readers may also view select video from the study on the National Institute of Neurological Disorders and Stroke YouTube channel, https://www.youtube.com/watch?v=QRt8QCx3BCo.

The formulation of the neuroprosthetics community as "sometimes vituperative" appears in a blog post by Gary Stix titled "Paralyzed Patient Swills Coffee by Issuing Thought Commands to a Robot" on the *Scientific American Observations* blog, http://blogs.scientificamerican.com/observations/2012/05/16/paralyzed-patient-swills-coffee-by-issuing-thought-commands-to-a-robot/.

9. THE REDEEMER

Scheuermann's e-book, *Sharp as a Cucumber* (Amazon Digital Services, 2012), is based on one of the plots she concocted when she worked as a murder-mystery party planner.

10. BLIND SPOTS

My discussion of the intellectual history regarding sensory perception is deeply indebted to the work of the neuroscientist David Eagleman. His book *Incognito* (New York: Vintage, 2012) contains a lucid account of the relationship between the work of the Scottish neurologist Charles Bell and the German physiologist Johannes Peter Müller. My discussion of the scientists' evolving thoughts on sensory perception draws from Eagleman's insights, which he presents succinctly and with great panache. My reading of these two scientists' work would not be possible without Stanley Finger's marvelous *Origins of Neuroscience* (New York: Oxford University Press, 2001), from which I draw scientific and biographical information as well as quotations for both scientists.

I am indebted to Eagleman, once again, for my characterization of how the neuroscientist Donald MacKay's work challenged prevailing theories on sensory perception in the mid-twentieth century and the implications it held for how the brain actively constructs vision. (Indeed, it was Eagleman who first taught me the trick to recognize the blind spot!)

Similarly, my discussion of how the brain creates vision is deeply influenced by the work and writings of V. S. Ramachandran. His book *The Tell-Tale Brain* (New York: W. W. Norton, 2012) informs my understanding of how the brain perceives optical illusions like the cube on page 203 and the picture on page 205.

Similarly, I draw on both Eagleman's and Ramachandran's writings, which dovetail nicely with current BCI research into motor planning, in my discussion of how the brain builds models of the physical world.

In addition to Eagleman's and Ramachandran's work on perception, I found the neuroscientist Sebastian Seung's book *Connectome* (New York: Mariner Books, 2013) particularly helpful for my discussion of how the so-called Jennifer Aniston neuron relates to the motor system's symbolic representation of the physical world.

11. FROZEN MIRRORS

Again, readers interested in Rizzolatti's original research should consult Pellegrino et al., "Understanding Motor Events." Rizzolatti et al., "Premotor Cortex and the Recognition of Motor Actions," is also helpful. But perhaps Gallese et al., "Action Recognition in the Premotor Cortex," gives the fullest account of this early work.

I draw on Iacoboni's *Mirroring People* in my discussion of how mirror and canonical neurons may relate to the motor system's symbolic construction of the physical realm. His theories, while not unique, complement the work of Schwartz and other BCI researchers as they explore theories of embodied cognition.

Readers interested in learning more about the Dutch study should consult Evelyne Kohler et al., "Hearing Sounds, Understanding Actions: Action Representation in Mirror Neurons," *Science* 297, no. 5582 (August 2002): 846–48.

For an academic account of the Italian team's research into varying neural activations for similar physical gestures, see Leonardo Fogassi et al., "Parietal Lobe: From Action Organization to Intention Understanding," *Science* 308, no. 5722 (April 2005): 662–67.

12. PIANO MAN

For Schwartz's account of the experiment with Scheuermann, see Jennifer L. Collinger et al., "High-Performance Neuroprosthetic Control by an Individual with Tetraplegia," *Lancet* 381, no. 9866 (February 2013): 557–64.

ACKNOWLEDGMENTS

It is with deep gratitude that I thank the many people who contributed to the completion of this book. I'm especially indebted to Jan Scheuermann, the Brookman family, and Tim Hemmes, who shared such intimate details of their lives with me. Eric Leuthardt, Andrew Schwartz, Jon Donoghue, Dan Moran, Leigh Hochberg, Gerwin Schalk, and Miguel Nicolelis were all extremely generous with their time, allowing me into their labs, sitting for countless hours of interviews, and fielding my oddball questions. I really couldn't have done it without you.

I am also indebted to a host of other scientists, whose names do not appear, or appear only briefly, in the book. In particular, I would like to thank Mohit Sharma, Nick Szrama, David Bundy, Jeanette Stoney, Jesse Wheeler, Mrinal Pahwa, Edward Hogan, Theodore Berger, Richard Normann, Greg Mark, Sliman Bensmaia, Rob Gaunt, Michael Boninger, Jennifer Collinger, Wei Wang, Anita Srikameswaran, Mikhail Lebedev, Laura Oliveira, Susan Halkiotis, Kevin Warwick, Philip Troyk, Allen Ravitz, Mike McLoughlin, Robert Armiger, Tan Le, Philip Kennedy, Brian Mech, Gislin Dagnelie, Christoph Guger, and Kathleen Snodgrass.

I am particularly grateful to the Alicia Patterson Foundation, which believed in the project early on and whose support was critical to its completion.

Special thanks also go to David and Anne Kendall, Tom and Sarah Wright, and Carl and Caryn Wright, who were all such gracious hosts during my travels.

This book relies on the hard work of a number of other journalists and scientists, whose writings and research I consulted throughout the project. In particular, I'd like to thank David Eagleman, V. S. Ramachandran, Antonio Damasio, Andy Clark, Marco Iacoboni, Sebastian Seung, Stanley Finger, Adam Piore, Jonathan Moreno, Jon Mukand, Richard Martin, Michael Belfiore, Daniel Wolpert, David Sloan Wilson, Mark Pagel, Alexis Madrigal, Tony Judt, Jessica Benko, and Tracey O'Shaughnessy, whose deep research, graceful writings, and fascinating insight opened the field to a novice like me.

Kathy Robbins and David Halpern are brilliant agents who offered constant support and expert guidance. Peng Shepherd was unfailingly patient, often anticipating many of my questions before they occurred to me. My deepest thanks goes to my editor, Eric Chinski, whose early suggestions and astute reading helped form the book's essential shape, bringing tremendous improvements to the manuscript.

This book would never have come into existence without the love and support of my friends and family. I'm particularly grateful to Ellen and Durb Curlee, who made sure I got away from the desk, recruiting me for beaver duties; Harper Barnes, who kept me focused; Chad Garrison, who offered me a welcome reprieve; Walker Gaffney, who livened up Squirrel Palace; Cale Bradford, who offered early support; and Joshua Correll—that bottle of Scotch is just about empty! My sister, Joscelyn, made sure I stayed grounded. My father and stepmother were powerful role models, great critics, and a constant source of encouragement. My mother and stepfather offered love and guidance. Finally, to my dearest, Allison: thank you for your patience, support, and enduring love.

INDEX

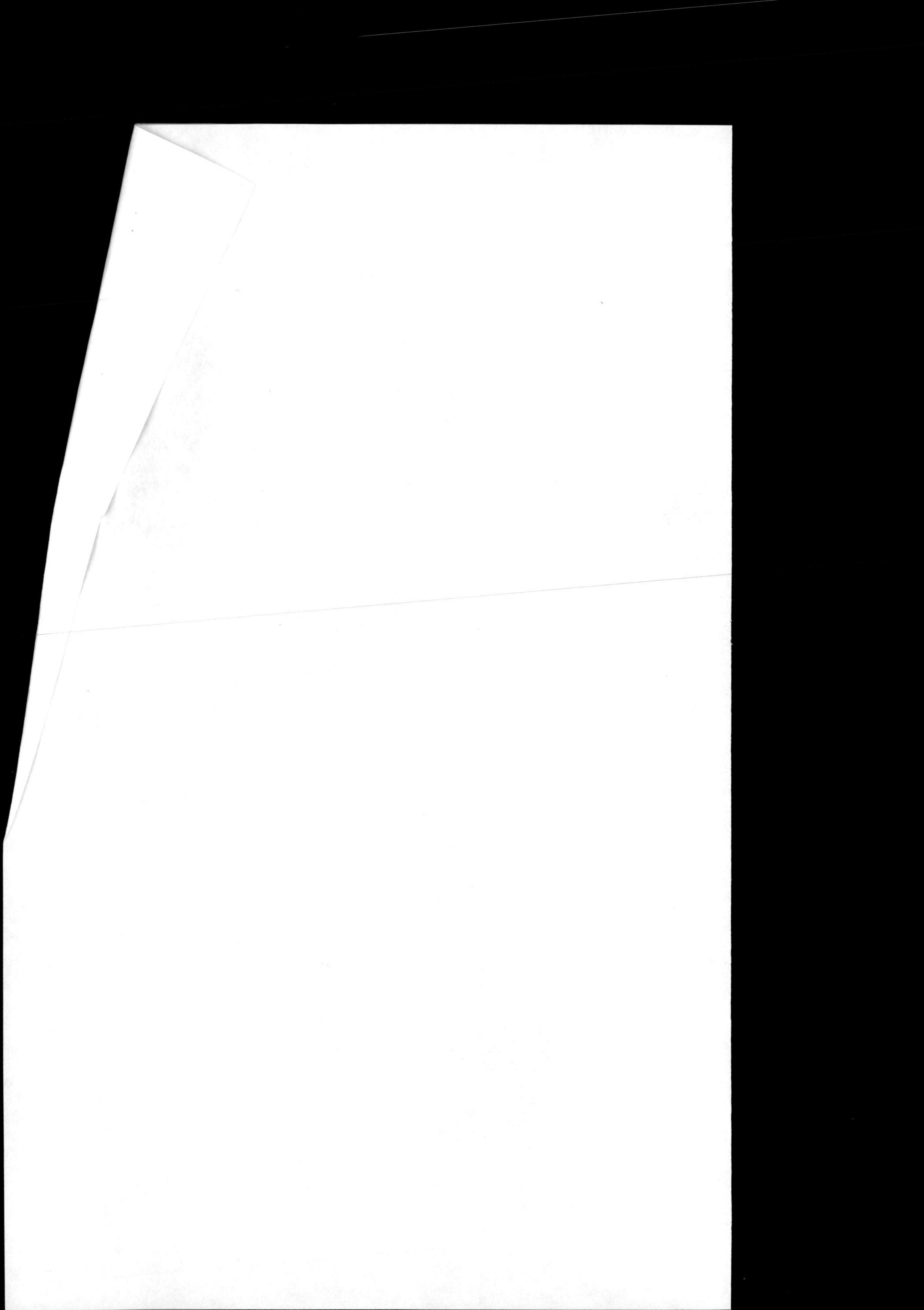